The Logic of Machines and Structures

The Logic of Machines and Structures

Paul Sandori
Department of Architecture
University of Toronto

QA
821
.S18
1982

1807 1982
175 YEARS OF PUBLISHING

A Wiley-Interscience Publication
JOHN WILEY & SONS
New York   Chichester   Brisbane   Toronto   Singapore

Library of Congress Cataloging in Publication Data:

Sandori, Paul, 1937–
    The logic of machines and structures.

    "A Wiley-Interscience publication."
    Bibliography: p.
    Includes index.
    1. Statics. I. Title.

QA821.S18   531'.12   81-19743
ISBN 0-471-86397-1   AACR2
ISBN 0-471-86193-6 (pbk.)

Printed in the United States of America

10 9 8 7 6 5 4 3 2 1

pd
6-24-83

To Petra and Mark

Preface

We regard a phenomenon as explained when we discover in it known simpler phenomena.   Ernst Mach

The principles governing the balance of forces on machines and structures are extremely simple and based on common, everyday experiences. Complex and even paradoxical devices can be reduced to these basic facts by a very straightforward process of logical reasoning.

Because of this simplicity of the principles and their inherent logical framework it was possible very early to condense our experience with simple machines and structures into a mathematical form—the science of mechanics—which has ever since served as a model to every other branch of science. Unfortunately, the mathematical form obscures the simplicity of mechanics for people who are not familiar with mathematics.

This book is an attempt to expose the principles of statics in their original simplicity and present it as an exercise in logic, a reasoning process without the mathematical shorthand. In this way, people with no mathematical background can understand machines and structures as Mach defined understanding: in terms of simple, familiar phenomena. At the same time, since the logic of mechanics is carefully preserved, it should also help students of engineering and architecture who are embarking on a modern mathematical course to become familiar with the thinking underlying the mathematical methods.

My second objective is to restore enjoyment to the study of the subject. Today essentially an engineering tool, mechanics was in its early days considered "the most noble of arts." For centuries some of the most penetrating minds expended their best efforts in its development; these brilliant early insights and intuitions are presented in this book to show how the modern science of mechanics has developed from them. I hope this approach will attract more high school students to the study of technical subjects and also appeal to those people who are interested in the history of human endeavor to explore and harness the forces of nature.

PAUL SANDORI

Toronto, Ontario
January 1982

Acknowledgments

This book has evolved over a number of years as a course given to my students in the Department of Architecture, University of Toronto. It would never have happened had they not shown interest and appreciation. I am grateful to the staff and students of the Institute for the History of Science and Technology who listened to my first ideas and discussed them with me and to Professor P. M. Wright of the Department of Civil Engineering, University of Toronto, who read the manuscript. Special thanks to my former students Alison McKenzie and Ian Petroff for their comments and encouragement and to my friend Ted Jeruzalski for his excellent freehand sketches. Finally, I would like to thank the editors and production personnel at John Wiley & Sons for their invaluable help in producing the book.

P. S.

Contents

Machines and Structures

Things are simpler than they seem

THE INTUITIVE APPROACH

The man changing the wheel on his wagon (Fig. 1.1) is faced with a problem: before he can remove the wheel, he has to raise the heavy vehicle, but muscular strength is not sufficient for the job. He solves the problem by means of a familiar device—a *lever*. A more sophisticated version of the lever is standard equipment on most cars today and, in even more complex forms, it is part of various machines used to exert enormous forces and pressures in industry.

The modest lever is a so-called *simple machine* and does the same job as its complex derivatives: it multiplies the effort of the human arm to achieve something that, unaided, the arm could not possibly do. The advance of our civilization started with devices like that.

Once propped up the lever becomes a structure. It is no longer acting to raise the load. It is fixed, and it holds the load up against the pull of gravity. We should now call it a *beam* spanning its supports on the ground and on the prop. Lever or beam, it is a simple enough device, familiar and easy to understand in its working. Let us now look at something less familiar.

The bridge over the Firth of Forth in Scotland (Fig. 1.2), completed in 1890, seemed an amazing feat of engineering at the time of its construction. Its twin spans of over 520 m each not only were the longest in the world but also had the strength and rigidity to support tremendous train and wind loads. Its huge size dwarfed anything built before: nearly 60 thousand tonnes of steel went into the bridge, held together by six and a half million rivets!

The bridge has been likened to the pyramids of Egypt, because of its

Figure 1.1   The lever—a machine and a structure.

monumental size and its significance as a landmark. People have been flocking to admire it ever since it was built. On the other hand, because of its unusual profile, it has also been likened to a dinosaur and a well-known contemporary aesthete, *William Morris*, called it "the supremest specimen of all ugliness."

Leaving aside aesthetics, let us try to answer a much more down-to-earth question: how does this bridge support the loads on it? What does what in the structure? What are the engineering reasons that dictated the unusual shape? At first glance, the answers to these questions may seem very hard for a nonspecialist to find, but this is not so. Figure 1.3, a photograph taken from a contemporary publication, contains an extensive structural analysis of the bridge based entirely on intuitive knowledge.

As can be seen from the diagram on the wall behind the live model, only one of the twin spans is shown. The man sitting on the stick in the middle of the model represents the load on the span, a train. The stick is the central girder between two piers of the bridge, called the suspended girder. The supporting structure projects outward from the piers on either side and is represented by the arms of the men sitting on the chairs and by the sticks held in their hands at one end while the other end is butted against the chair. Such a structure projecting over the support is called a *cantilever*.

Figure 1.2 The Forth Bridge, completed in 1890. Its twin spans were the longest in the world for almost 30 years.

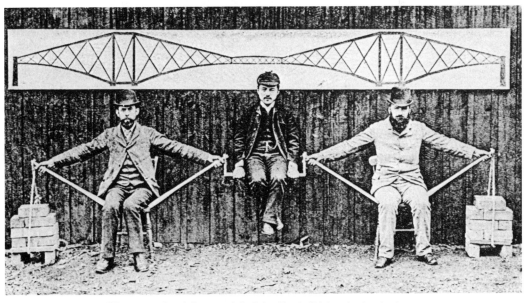

Figure 1.3 A live model of the Forth Bridge, by its designer.

The load from the suspended girder is transmitted to the cantilevers. It tends to make the pier overturn but, on the opposite side of the support, it is balanced by a counterweight. In the model, the counterweight is a stack of bricks; in the bridge, it is in the masonry towers at either end of the bridge.

What happens inside the structure is equally easy to understand intuitively. The arms of the men and the sticks work together in resisting the load. The load pulls on the arms and pushes on the sticks. The arms are in *tension*; the sticks are in *compression*. In a structure, the compression elements—the sticks—are called *struts* and the components in tension—the arms—are called *ties*. Another part of the structure that is evidently in compression are the bodies of the two men. They have to make an effort to hold up their shoulders just as if the load had been placed directly on them.

Looking more closely at the bridge itself (Fig. 1.4), we can easily differentiate between ties and struts. The ties are transparent, light, delicate-looking lattices; the struts huge tubes. Experience provides a the reason for that too. A long, slender piece of steel or wood can take a large pull in tension but far less in compression. It simply bends out under a comparatively small push. This type of structural behavior is called *buckling*. A tube is a particularly efficient structural shape when it comes to resisting buckling.

The stick on which the man in the middle is sitting is neither a tie nor a strut. Straight when unloaded, it *bends* under the load; it works as a *beam* just as the propped lever did in the first example. Bending is far more damaging to a structural member than either tension or compression. If we wanted to break a stick, we would not for a second entertain the idea of breaking it by pulling or pushing at the ends to produce tension or compression. The easiest way to break it is by bending it. Consequently, in the bridge we notice that the suspended girder is kept comparatively short and, moreover, it is made up of struts and ties, just like the rest of the bridge. Bending is, somehow, resolved into tension and compression. Such a structure is called a *truss*.

In the following chapters, we develop a method of analyzing structures and machines in terms of a small number of basic principles. This method—the result of centuries of thought and experience—reinforces intuitive and practical understanding of how machines and structures work while at the same time making us independent of it.

THE ART OF WEIGHING

The method of analyzing machines and structures that we are looking for originated many centuries ago in attempts to understand devices like the lever which, as if by magic, multiplied the effort of the person using them. (Structures did not attract much attention: there was nothing mysterious

Figure 1.4 Transparent lattices and huge tubes—the tension and compression elements of the bridge.

about the way a beam supports its load and, if one broke, it was simply replaced by a bigger one.) For the person seeking a logical explanation of machines and structures rather than a detailed engineering analysis, it is profitable to go back to the origins and start where the early investigators started: with simple, basic problems involving familiar devices. Only those texts that particularly suit the purpose of this book are discussed here.

In 1586 a Dutch engineer and mathematician, *Simon Stevin* (Fig. 1.5), published *The Elements of the Art of Weighing*, a book that dealt mainly with the balancing of loads. There are several reasons for choosing Stevin's work as our starting point. Stevin had an uncommon ability to combine theory and practice, very probably due to the requirements of his job: he was an engineer, a quartermaster of the army, and an inspector of dikes and waterways in the Netherlands while at the same time acting as mathematics and science tutor to Maurice of Nassau, Prince of Orange. Stevin must have been

Figure 1.5   Simon Stevin (1548–1620).

an excellent teacher for he had a genius for deriving his explanations from axioms (self-evident facts) without recourse to any previous knowledge of mathematics. Moreover, in his discussions he instinctively chose the path that, in an organized and developed form, constitutes the modern method of analysis.

Stevin's work on the art or science of weighing was paralleled by the writings of many scientists before and after him. We shall borrow from the work of just one of them: the greatest Italian scientist of the period, *Galileo Galilei* (Fig. 1.6). After his trial and condemnation by the Roman Inquisition in 1633 for his heretical views regarding the motion of the planets, Galileo was forced to retire. He was forbidden to express his opinions "in writing or speech or in any other way, on the movement of the earth and the immobility of the sun." Instead, he managed in 1638 to have a book published called *Two New Sciences.* One of the new sciences dealt with bodies in accelerated motion; the other explored the resistance to fracture of beams. For the first time structures received a scientific treatment.

At first glance, water, air and other fluids appear to be substances that have little to do with machines and structures, either in their application or in their behavior. Appearances are misleading on both counts. Stevin wrote a book on buoyancy and on the effects of the pressure of water, coming to some startling conclusions. He laid down the fundamentals of the subject but his work did not have much influence until the thread was picked up again by the French mathematician and philosopher *Blaise Pascal,* who in 1663 published his *Treatise on the Equilibrium of Liquids and the Weight of the Mass of the Air.* Pascal did not really go much further than Stevin regarding water pressure; what he did though was to produce many ingenious proofs and experiments covering the subject more systematically and clearly. In one respect he made a significant advance: he made use of his understanding of the behavior of liquids under pressure to invent a new machine—the hydraulic machine.

The pressure of the air was not touched upon at all by Stevin. The pioneering work in this respect was done by Galileo's pupil *Torricelli* and further developed by Pascal (as the title of his book indicates). Pascal showed that a large group of phenomena thought to be caused by an abhorrence of the vacuum, attributed to Nature ever since *Aristotle,* can be explained in terms of pressure of the air, closely resembling the effects of the pressure of water. At approximately the same time the mayor of Magdeburg in Germany, *Otto von Guericke* (Fig. 1.7), made many practical experiments involving air pressure, some with really spectacular results. These stimulated other scientists and eventually led to the invention of the steam engine that revolutionized industry a century later.

Figure 1.6   Galileo Galilei (1564–1642), aged 78.

Finally, we shall take Stevin's work on the workings of another labor-saving device, the pulley, as the starting point of a different approach to problems involving machines and structures. The results obtained in this

Figure 1.7   Otto von Guericke, the mayor of Magdeburg (1602–1686)

way will, of course, be the same answers as before but seen from a new point of view and providing new insights. Stevin himself is often credited with developing this alternative. Actually, he did his best to suppress it as erroneous.

The objective of this infusion of historical examples, as noted at the start, is an intuitively and rationally understandable logic of structures and machines. However, there is another reason. The originality and sheer brilliance of the reasoning contained in the work of people like Stevin, Galileo, and Pascal is in itself enjoyable and worth study as part of our heritage in the fields of science and philosophy.

FORCES AND EQUILIBRIUM

Let us look at how Stevin applied his *Art of Weighing* to a problem—a treadmill crane that enjoyed widespread popularity in sixteenth-century Europe just as it did in Antiquity (Fig. 1.8). One or several men would make the large wheel of the crane turn by forever climbing up inside; this would wind a rope around the axle *EF* and cause the load *H*, hanging on the rope, to be lifted. How much load can a single man lift in this way?

In the theoretical part of his *Art of Weighing* Stevin deduces from a very basic axiom what was then known as the *law of the lever*. The load *H* on the short arm of the lever and the effort *E* applied at the end of the long arm (Fig. 1.9) stand in a certain ratio to each other which depends on the relative lengths of the lever arms *AG* and *GF*. These ratios are given by the law of the lever (which we examine in Chapter 2). Stevin's analysis of the machine consists of showing that it functions as a simple bent lever. As the man climbs toward point *A*, he pulls down the long lever arm *AG*. The short arm

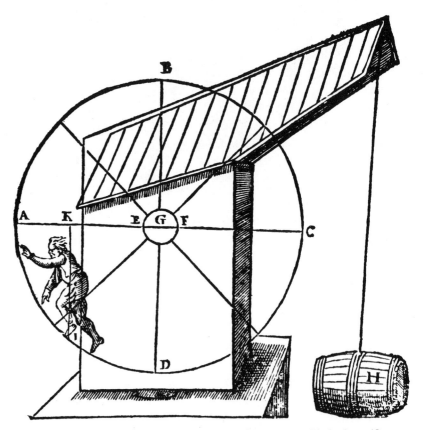

Figure 1.8   A Dutch treadmill crane. Can the man lift the barrel?

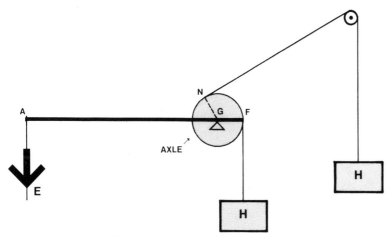

Figure 1.9 The crane simplified: it acts as a lever.

of the lever is represented by the radius *GN* of the axle on which is exerted the pull of the load. Anybody who has mastered the art of weighing and knows the law of the lever can now figure out easily the answer to the problem of the crane.

The art of weighing does not exist any more. It has been replaced by the science of *mechanics*. The concept of weight has lost its special significance. Weight is the pull of gravity and is in no way different from the pull of a rope or the push exerted by a man. In mechanics, any push or pull, however caused, is called *force*. The weight of the load at one end of the lever is a force, just as the push of the man is a force at the other end.

Consider another illustration from Stevin's book, reproduced in Fig. 1.10. Stevin employs a variety of devices to show the pulls exerted on a prism. A string passing over a pulley with a load at the far end exerts an oblique pull. For an upward pull the string is attached to a balance with a load hanging at its opposite end—then the upward pull of the string is equal to the weight of the load. Whenever required, a helpful hand reaches down from the sky to hold any loose ends.

The modern equivalent of this illustration, seen in Fig. 1.11, is less vivid but much simpler. All the props, including the pyramid support, have been eliminated and replaced by arrows indicating forces following very specific rules which we shall see in a later chapter. This is one of the problems with modern mechanics: it is simple almost to the point of being dull.

The principles of mechanics are no longer concerned with the law of the lever or some similar device. Mechanics deals only with *bodies*, and all bodies, regardless of their shape, size, or function, behave in the same way under the action of forces. A lever is a body which, in practice, happens to be a simple machine; in mechanics, this does not entitle it to any special treatment. The old law is simple enough but we reduce it and all the other

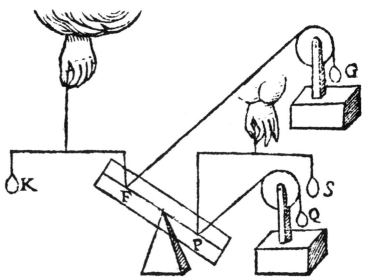

Figure 1.10 Pulleys and helpful hands from the sky: a problem in the art of weighing.

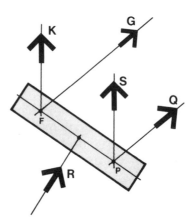

Figure 1.11 Modern representation of the same problem. All props have been eliminated.

similar laws and rules to a small number of even more basic principles which concern forces and bodies. Armed with these principles we proceed in pretty much the same way as Stevin did in his treatment of the crane. We discover these same simple principles at work in all the complex problems that we may encounter.

The forces acting on a body are either balanced or not. If they are not balanced, the body will be subjected to accelerated motion. The branch of mechanics that deals with accelerated motion is called *dynamics*. It was developed by *Isaac Newton* (following some pioneering work by Galileo) in

Example   13

his famous book *Mathematical Principles of Natural Philosophy*, published in 1687. The branch of mechanics that investigates forces in a state of balance, or *equilibrium*, is called *statics*. This is the science that replaces the ancient art of weighing and is the subject of this book.

The world of statics is inhabited only by rigid bodies and forces. Everything in it can be measured in terms of two kinds of units: the units of force, *newtons*, and the units of length, *meters*. These are the units of the metric International System of Units (abbreviated SI), adopted in 1960 by a world conference on weights and measures. For uniformity we use SI units even in examples that go back to the time long before the metric system was invented. The simplicity of statics makes it ideally suited for a mathematical treatment and, in fact, mechanics was the first science that was put in a mathematical framework with all the other sciences later trying to approach this ideal.

Over the centuries, the mathematical side of statics has become extremely well adapted to the job and streamlined to perfection. In fact, many machines and structures are now possible not only because of the advances in technology but also because they can be analyzed mathematically. The CN Tower in Toronto (Fig. 1.12), the tallest freestanding structure in the world, is a good example: its design was performed to a large extent with the help of computers. However, this is a separate development. We can understand the structure of buildings such as the CN Tower perfectly well without a mathematical analysis. Statics offers a quick and reliable method of understanding the effect of forces on bodies, or, in practical terms, a *logic of machines and structures*.

EXAMPLE

We can easily understand the structural action of the Firth of Forth bridge once we see the live model of Fig. 1.3. In fact, we can feel the work required of the various structural components of the bridge in our muscles and bones. This is not surprising. After all, the job that structures perform on a large scale—that of carrying and balancing loads—is on a smaller scale an everyday experience for humans and animals.

The skeleton of the bison seen in Fig. 1.13 bears a remarkable similarity to the structure of the Forth bridge, superimposed in outline on the skeleton. The bones are the compression elements while the muscles and tendons of the animal, working in tension, complete the structure. We can safely assume that the muscles in the neck and back of the bison follow the direction of the ties in the bridge since they perform the same function.

The differences between the two structures stem from the need for flexi-

Figure 1.12 The world's tallest free-standing structure: the CN Tower in Toronto, 550 meters.

Example   15

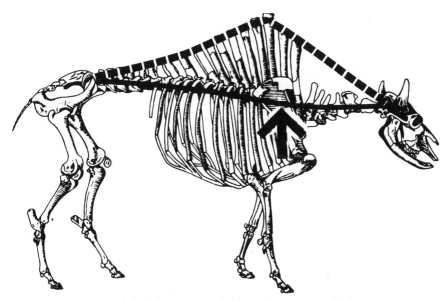

Figure 1.13   Not unlike a bridge: the structure of a bison.

bility in the animal (something the engineer is at pains to avoid in a bridge) and the dynamic requirements of quadrupedal motion (totally absent in a bridge, of course). This last requirement, for example, accounts for the fact that most of the structure and therefore weight of the body is concentrated on the forelegs. The cantilever is so balanced over its support that, while anchored at the back, it transmits very little of its weight to the hind legs which serve to push the body forward. There are no bracing struts in the flexible neck which, supporting the heavy head of the animal, is constructed like the arm-and-stick arrangement in the live model of the bridge.

The Lever

A paradox of nature explained

THE LAW OF THE LEVER

Aristotle considered the lever a paradox of nature. A small force applied to one end of it, a long way from the point of support (called the *fulcrum*), overpowers a much larger force applied near the support. How is that possible?

Ever since Antiquity, philosophers and mathematicians felt challenged to discover the relationship between the forces on the lever and their distances from the fulcrum. The first to come up with the right answer and rationally prove it (as far as we know) was the Greek philosopher *Archimedes* in the third century B.C. However, we shall start with the arguments presented by his ardent follower, *Simon Stevin*. In his book *The Elements of the Art of Weighing*, published in 1586, Stevin begins with an axiom: a uniform rectangular beam *AB* (Fig. 2.1) will balance if suspended in line with its midpoint *T*. This midpoint is called its *center of gravity*.

The beam is shown consisting of six identical units. If we make a cut along line *EF* (Fig. 2.2), the beam will be divided into two unequal parts. The part to the left of line *EF* is 4 units long, and the part to the right is 2 units long. Now imagine the two parts each suspended from its own middle point and attached to a lever (Fig. 2.3). Assume that neither of the threads from which the two parts hang nor the lever has mass. The lever serves only one function: it supports the original beam (now cut in two) in its original position, that is, hanging from a thread in line with its midpoint. The center of gravity of the body was not displaced sideways by the cut or by moving the

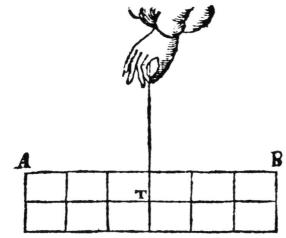

Figure 2.1   A beam balanced about its center of gravity.

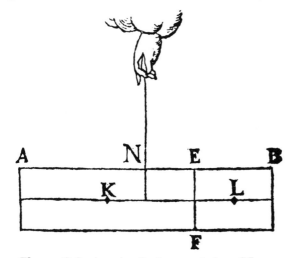

Figure 2.2   Imagine the beam cut along *EF*.

two parts up or down, so the lever with the two portions of the beam hanging from it must balance.

From the illustration we see that the distances of the smaller and larger parts from the fulcrum are 1 unit (distance *KN*) and 2 units (distance *NL*), respectively. The larger part weighs 4 units, the smaller part 2 units. Hence the weight forces on the lever and their lever arms are in the ratio

$$4 : 2 = 2 : 1$$

and since the cut *EF* could have been made anywhere with similar results (rigorously proven by Stevin), it can be said that generally two loads on a

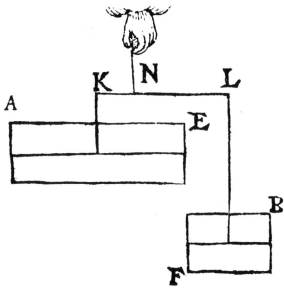

Figure 2.3 The two portions of the beam are still balanced about the center of gravity of the entire beam.

lever will balance if their magnitudes are inversely proportional to the lengths of their lever arms:

$$A : B = NL : NK$$

This is the *law of the lever.*

Instead of the two portions of the beam (A and B), any two masses of the same weight as the blocks A and B can be suspended from the lever (Fig. 2.4). The lever will remain in equilibrium because the pulls at its ends and the lever arms have not changed. Furthermore, one of the pulls, for example, pull B, can be exerted by hand. Then the smaller effort B will balance the larger pull of the load A. The minutest extra effort at B will, in the ideal circumstances always assumed in these arguments, suffice to lift the load.

Having thus arrived at the law of the lever, Stevin uses his knowledge to solve practical and theoretical problems. Some of his practical applications are not of much interest today. The question relating to Fig. 2.5, for instance, is the following. What is the pull that the soldier has to exert on his lance (a) if he carries just the lance and (b) if he also carries a poached chicken hanging at point K? We pass over this problem, even though the law of the lever is obviously applicable. In a later section we investigate the application of the principle to some simple machines based on the lever.

Stevin's theoretical considerations revolve around the rectangular beam and its equilibrium. In Fig. 2.6 the beam is shown with a load attached. If the beam weighs 6 units and the load is of the same weight, where should

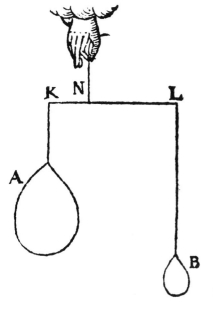

Figure 2.4 The law of the lever: $A : B = NL : NK$.

Figure 2.5 A practical application of the law of the lever—how to support a stolen chicken.

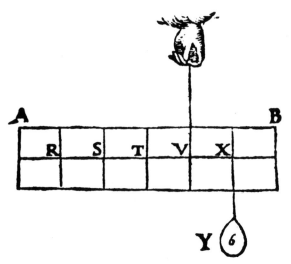

Figure 2.6   A theoretical application of the law of the lever.

the supporting string be attached to preserve the state of balance? The center of gravity of the beam, at point T, is where the weight force is applied while the other load hangs at point X. These two points can be considered the ends of an imaginary lever; since the loads on the lever are equal, the fulcrum should be the same distance from both ends, that is, at point V.

Figure 2.7 presents a slightly more complex problem. In addition to the weight of the beam itself, there are now two loads hanging from it, a load of 24 units having been added at point R. The weight of the beam and the load at point X can be replaced by an imaginary load of 12 units hanging at point

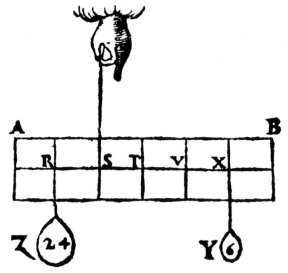

Figure 2.7   A slightly more complex theoretical problem.

V because, reasons Stevin, this is what we found we had to support at that point in the previous example. Now the new lever is the length *RV* and the fulcrum should be located at point *S* dividing the lever in the ratio 1:2 (the loads being in the ratio 24:12). Ingenious, but by no means straightforward. The following section outlines a much simpler approach.

THE MOMENT OF A FORCE

Stevin expressed the law of the lever in terms of the loads balanced at its ends, but it was understood that the law would be equally valid if the pull at one end or the other was exerted by hand. The loads were just a convenient way of representing a pull. Figure 2.8 shows a modernized version of Stevin's lever. In present-day statics, any push or pull, however caused, is called a *force* and represented simply by a line with an arrow on it.

The line (called the *line of action* of the force) and the arrow together indicate the *direction* of the force, thus replacing the string in Stevin's diagrams. It does not matter whether the pull of the force is exerted on a long or short string or even if we substitute a push for a pull. The effect of the force remains the same. Consequently, the line of action of a force can be extended indefinitely at either end, and the force arrow can be moved to any position on it. Figure 2.8 shows three equivalent positions of the arrow representing the force holding up the lever.

For a force to be fully defined we must also show how much pull it exerts. This *magnitude* of the force is marked beside the arrow in units of force called *newtons*. Of course, these are not the same units that Stevin used— Newton had not even been born when Stevin was writing his book. If the magnitude is not known it will be denoted by a letter, as in algebra. For

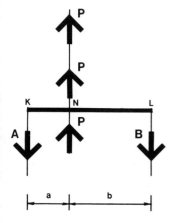

Figure 2.8 Modern representation of the lever. Force arrow **P** can slide along its line of action.

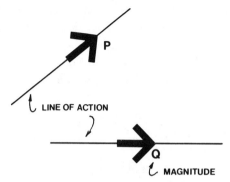

Figure 2.9   Two forces fully defined: direction and magnitude are all we need to know.

example, the letter P denotes the magnitude of the support force which is not yet known.

These two characteristics, direction and magnitude, are all we need to know about a force. On the other hand, both direction and magnitude must be known in order to define a force. Force is a *vector* as opposed to, say, mass, which has magnitude but no direction. Mass, and any similar quantity that has only magnitude, is called a *scalar*. When we talk about a force in this book, a boldface capital letter is used to denote the force while the same letter, in italic type, will mean the magnitude of that particular force. For example, the two forces in Fig. 2.9 have the magnitudes P and Q, respectively, marked on the drawing in regular letters. In the text, if we want to talk about the forces and not just their magnitudes, we must use the boldface letters **P** and **Q.**

The law of the lever (referring again to Fig. 2.8) states that

$$A : B = b : a$$

For 2000 years since Archimedes all writers (including Stevin) belabored this inverse proportionality between forces and lever arms, for the simple reason that the branch of mathematics they were applying was geometry. That was also the only branch of mathematics sufficiently developed at the time. We now convert this ratio into a mathematically equivalent expression, more suited to modern algebra:

$$A \times a = B \times b$$

The product (magnitude of force × lever arm) is called the *moment* of the force, or *torque,* and is a measure of its turning effect. The units of moment are those of force and length combined—newton-meters. The lever is bal-

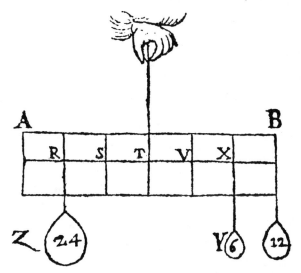

Figure 2.10   A beam in equilibrium. The resultant moment of the forces is zero.

anced on the fulcrum when the moments of the two loads are equal and opposite: the clockwise moment of force **B** is balanced by the counterclockwise moment of force **A**. The two moments cancel each other out.

The concept of moment, thus liberated from the lever where it originated, simplifies many things. For example, is the beam shown in Fig. 2.10 in equilibrium? There is no need to look for a lever to find out. The clockwise moments about the support at point T are

$$6 \times 2 + 12 \times 3 = 48 \text{ newton-meters}$$

The counterclockwise moment is produced by a force of 24 newtons and is equal to the total clockwise moment:

$$24 \times 2 = 48 \text{ newton-meters}$$

Therefore the beam is in equilibrium. The weight force acting on the beam at its center of gravity was not taken into account because it has no moment about the fulcrum T: the two points coincide and the lever arm of the force is zero.

This can be refined further. Moments can turn the body on which they act either in a clockwise or in a counterclockwise sense. We distinguish between the two by giving a positive sign to clockwise moments and a negative sign to those acting in the opposite sense. Then we can add or subtract moments and express the law of the lever in yet another way. The sum total of the moments acting on a body is the *resultant moment*. The lever is balanced when the resultant moment about the fulcrum is zero. Thus in Fig. 2.10 we have

$$-24 \times 2 + 6 \times 2 + 12 \times 3 = 0$$

The lever is in equilibrium. Suppose we add another 2 newtons to the force acting at point X. Then the resultant moment is

$$-24 \times 2 + 8 \times 2 + 12 \times 3 = 4 \text{ newton-meters}$$

and the body will turn clockwise.

THE MOMENT ARM

Archimedes and Stevin and most of the writers between considered the lever as a horizontal bar with forces on it pulling down vertically at different distances from the fulcrum. The principle of moments is much more general. Let us take as an example the problem seen in Fig. 2.11, where Stevin attempts to answer the following question. Suppose we already know the vertical force **I** that holds the beam in equilibrium. What must be the magnitude of the force **G** on the oblique string to replace the vertical force?

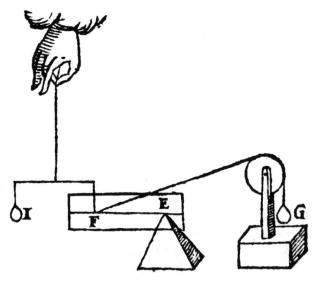

Figure 2.11   Can the oblique string replace the vertical pull?

Figure 2.12   The wheel-and-axle machine.

Do we need more or less force in the oblique string? Can the oblique force replace the vertical one at all? For Stevin, this was no longer within the boundaries of the law of the lever. For oblique (nonvertical) forces he made use of another principle that we examine in Chapter 3, but the problem can be very easily solved once we establish the exact meaning of the concept of lever (or moment) arm. This concept, on a practical level, is intuitively familiar; if not, it can be discovered quickly through experience with devices like the one shown in Fig. 2.12. It is called a wheel-and-axle machine except that in this instance the wheel consists only of spokes.

Suppose the load **W** on the axle is balanced by a counterweight of magnitude P hanging at the end C of a horizontal spoke (Fig. 2.13). The moment of the force

$$M = P \times AC$$

makes the wheel turn against the resistance of the rope coiled on the axle and pulled down by the load. It is intuitively quite clear that the same

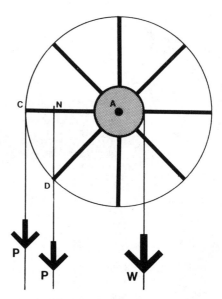

Figure 2.13 The physical lever arm is not the moment arm of the force. The turning effect of force **P** is reduced when it acts at point *D*.

vertical force **P** would not produce the same moment pulling down at point *D*. Indeed, if we turn the wheel a bit more so that the spoke is vertical and falls in line with the force, there will be no turning effect at all. Since the moment depends only on the magnitude of the force and its lever arm and since the magnitude has not changed, it follows that the moment arm must have become shorter even though the length of the lever (that is, of the spoke) has remained the same. The moment arm of a force is therefore not the physical lever arm on which it acts.

Purely on the basis of experience the moment arm of a force is defined as the shortest distance between the line of action of the force and the fulcrum. This distance is obtained by dropping a perpendicular line from the fulcrum to the line of action of the force. The turning effect of force **P** is the same as if it was hanging from point *N* instead of *D*. In other words, the moment of the force is equal to

$$M = P \times AN$$

The horizontal distance *AN*, at right angles to the vertical line of action of the force, is the real moment arm of the force. To produce the same turning effect as at point *C*, the push applied at point *D* would have to be increased to compensate for the reduced moment arm. A man pushing at point *D* would, to get maximum benefit for his effort, instinctively push in a direction at right angles to the spoke. Then the moment arm of the force is equal to the full length of the physical lever arm, the spoke.

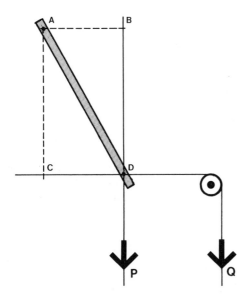

Figure 2.14 The moment arm of force **P** is the distance *AB*, that of **Q** is the distance *AC*. Therefore: *P* × *AB* = *Q* × *AC*.

Long before Galileo, *Leonardo da Vinci* understood the concept of moment perfectly. In a sketch in his notebook (Fig. 2.14), he shows a rigid bar *AD* free to turn about a hinge at point *A*. At the end *D* of the bar hang two loads, **P** and **Q**. The string to which the load **Q** is attached passes over a pulley as shown. The moment arm of the two forces is not the length of the bar *AD*, says Leonardo, but the distance *AB* for the force **P** and the distance *AC* for the force **Q**. Note how the lines of action of the forces are extended beyond point *D* to find their moment arms. Note also that force **Q** is not the load hanging on the vertical end of the string but the horizontal pull of the string itself on the bar! The bar will be in equilibrium if

$$P \times AB = Q \times AC$$

Now, with the help of the concept of moment arm, we can solve Stevin's problem of Fig. 2.11 without too much trouble. The forces acting on the body are shown in Fig. 2.15. To replace force **I**, force **G** must have the same moment about support *E*, or

$$G \times NE = I \times FE$$

The moment arm *FE* (as measured in the diagram) is three times longer than moment arm *NE*. Therefore, force **G**, acting on the shorter moment arm, must be three times larger than force **I** to produce the same moment.

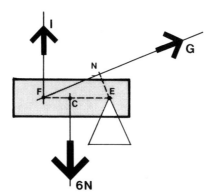

Figure 2.15 Solution of the problem of Fig. 2.11: $I \times FE = G \times NE$, therefore G must be three times the magnitude I.

The magnitude of force **I** required to resist the weight force of 6 newtons on the body is another problem that can be solved in the same way.

All reference to the lever has been gradually omitted. Any number of forces pulling on a body in arbitrary directions have no rotational effect about a fulcrum if the sum of their moments—their *resultant moment*—about that point is zero. This is no longer the law of the lever: this is a general condition of *rotational equilibrium*.

SIMPLE MACHINES

Any device that transfers a force from the point where it is applied to another point where it is used is a *machine*. The lever is a machine since the effort applied on one side of the fulcrum can lift a load on the other side. It does this in so uncomplicated a fashion that it cannot be further simplified. It is therefore called a *simple machine*.

Let us now look at some other simple machines whose working can best be explained by the principle of moments. They are no longer as familiar as they were in Stevin's time but, in various modified forms, they are an essential part of most complex modern machinery. With the growing scarcity of fuel they may once again become part of everyday life.

Useful though it is, the straight lever working on a fixed fulcrum has a severe limitation. It can only raise a load to a height above the fulcrum equal at the most to the length of the short arm. This difficulty is overcome by another simple machine, the wheel-and-axle arrangement (Fig. 2.16) that we have already seen. The load **I** hangs from a rope coiled on the axle. In Stevin's illustration the load is held in balance by a force **K**, represented by the counterweight hanging from the wheel.

At any moment, the device functions as a straight lever; the radius R of

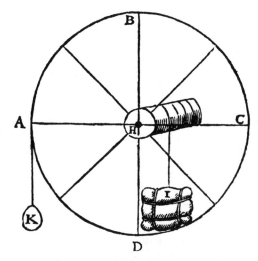

Figure 2.16   The wheel-and-axle machine.

the large wheel being the long moment arm of the force **K** and the radius r of the axle, the short arm on which hangs the load **I** (Fig. 2.17). As each radius moves out of position when the wheel is turned, the next takes its place. The wheel-and-axle machine can therefore be regarded as a continuous lever. Then from the condition of moment equilibrium

$$K \times R = I \times r$$

and

$$K = I \times r/R$$

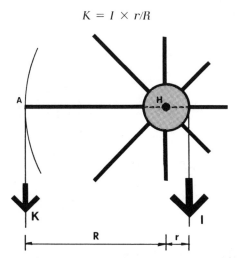

Figure 2.17   The wheel-and-axle analyzed in terms of moments of forces.

The smallest extra force will raise the load. A more sophisticated application of the principle is the treadmill crane seen in Fig. 1.8. The load to be lifted does not hang vertically down from the axle. Instead, the rope is carried up over a pulley and the short arm of the lever (radius r) is no longer horizontal. In every other respect the crane is the same as the wheel-and-axle. The energy crisis will probably have to become very severe before this particular machine experiences a revival.

Another simple machine in which we can discern the moment principle is the pulley. In its crudest form this is just a wheel with a grooved rim around which is passed a rope (Fig. 2.18). The rope is attached to a support at one end, while the effort **F** is applied at the other. The load **B** hangs from the axis of the wheel. When the pulley is stationary, the wheel is really superfluous: only the horizontal line ED is actually used and acts as a lever. Its fulcrum is at point D where the wheel is supported by the rope; the arm of the load is the radius r of the wheel while the arm of the effort is its diameter. Since the lever arm of force **F** is twice that of the load **B**, we

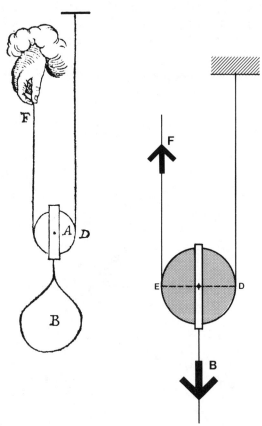

Figure 2.18   The movable pulley: in essence, it is a lever ED supported at D and pulled up at point E.

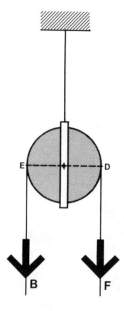

Figure 2.19 The fixed pulley does not multiply force. The lever *ED* is supported in the middle.

conclude that $F = B/2$. The weight of the pulley itself must also be lifted and should be added to the load.

As the load is raised by pulling on the rope the ratio of the moment arms is maintained while the machine rises with the load. For this reason, the pulley was called a traveling lever. It should be noted that only such a movable pulley doubles the lifting capacity of the effort. A different kind of lever is the fixed pulley (Fig. 2.19), so called because the axis of the pulley is attached to the support and acts as the fulcrum. The moment arms of the string forces on either side are the same in this case and so the pulls (the load and the effort) are also the same. All that this machine does is change the direction of the string force. It helps reduce the effort required to lift a load only when combined with one or more movable pulleys.

Sometimes the force that operates the lever is applied so that its moment arm is shorter than that of the load. For example, human and animal muscles attach close to the joints and work at a mechanical disadvantage: a large effort is required to produce a comparatively small force. On the plus side of this arrangement is the compactness of the body and the speed of movement: a short contraction of a muscle near a fulcrum produces a quick and longer movement at the far end of the lever.

A burrowing animal, such as a mole, has short limbs and muscles that act farther away from the joints—it moves slowly but it can develop relatively more power. On the other hand, a fast-moving animal, such as a horse, has long legs and muscles acting very close to the joints. The excavator in Fig. 2.20 requires speed and wide scope of movement as well as power. Consequently, it is designed so that the effort acts on the short arm of the lever but

Figure 2.20 A mechanical excavator. A huge force must be exerted on the short arm of the lever.

it also has "muscles" capable of exerting huge forces. "Muscles" in this case means the slender hydraulic cylinders—but that is another story. More about them in Chapter 5.

CENTERS OF GRAVITY

The starting point of the investigation of the lever was the axiom that a uniform beam will balance about a point called its center of gravity. The law of the lever and the concept of moment, in turn, are ideal tools to explore centers of gravity. Again we borrow from Simon Stevin some ideas on this subject but restrict our discussion to deal only with plane bodies of uniform thickness and density.

The center of gravity of a body can be found by experiment. Let us start with a triangle. Suspend the body from a point (Fig. 2.21). The triangle will come to rest under that point and, if displaced, return to exactly the same position every time. This is easily explained. Every body behaves as if its entire mass were concentrated at its center of gravity and the line of action of the weight force always passes through that point. When the triangle hangs from point A the center of gravity must be vertically under it or the force of gravity will have a moment about point A and rotate the body.

We conclude that in the triangle ABC the center of gravity will be located somewhere on the line AE. To find the actual position of the center we have to repeat the experiment from another point, point C perhaps, which will

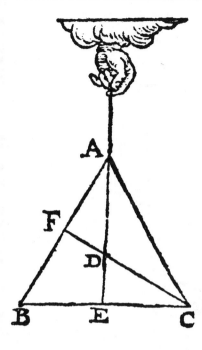

Figure 2.21   Finding by experiment the center of gravity of a triangular body.

give us line *CF.* Since the center must lie both on the line *AE* and on the line *CF*, it must be at point *D*, which is common to both lines. Figure 2.22 shows the location of the center of gravity *F* of a pentagon.

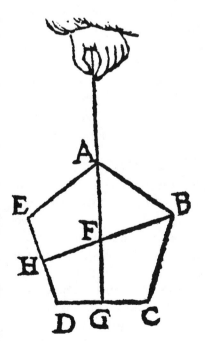

Figure 2.22   The center of gravity of a pentagon by experiment.

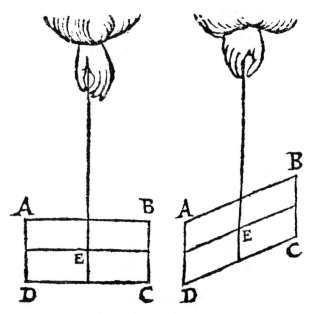

Figure 2.23   The center of gravity of a quadrangle.

For simple geometrical shapes there is no need for experiment. We can locate the center of gravity by reasoning. If a body has an axis of symmetry, like the triangle and like the rectangle in Fig. 2.23, the center of gravity will lie on the axis. This is self-evident for the horizontal rectangle. It follows that if the body possesses two axes of symmetry, the center is located at the intersection of the two axes: point *E* for our rectangle. The skew quadrangle beside it is symmetrical not about an axis but about its center of symmetry *E* because to every point such as *A* corresponds a point, *C* in this case, such that the line *AD* is divided into two equal parts by point *E*. A little reflection will show that the center of gravity must coincide with the center of symmetry.

Stevin finds the center of gravity of an asymmetrical triangle (Fig. 2.24) by reasoning as follows. Imagine three quadrangles inscribed in the triangle. The center of gravity of each quadrangle is at its center of symmetry. They all lie on line *AD*. Next suppose that the triangle is further subdivided into an infinite number of such quadrangles, each infinitely thin. Their centers of gravity will still lie on line *AD* but these quadrangles now completely fill the triangle. Therefore, the center of gravity of the triangle must also lie on line *AD*, which is called the median of the triangle. Using the modern method of the calculus, we would follow a very similar line of thought to arrive at the same result.

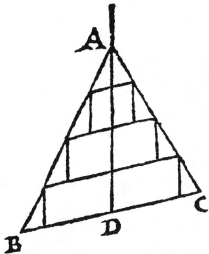

Figure 2.24 Quadrangles inscribed in a triangle. Its center of gravity must lie somewhere on the median *AD*.

Thus to locate the center of gravity *F* of any triangle (Fig. 2.25) we have to draw two medians (*AD* and *CE*). They both contain the center of gravity and therefore it must be at their intersection. Geometry further shows that the distance *FD* is one-third of the length *AD*, and *FE* is one-third of *EC*.

The center of gravity of more complex shapes can be found by subdividing them into simpler components. Stevin shows how to find the center of gravity *N* of an irregular quadrangle by cutting it into two triangles or of an irregular pentagon by dividing it into a quadrangle and a triangle, and so on. However, let us use a different example: a flat body of uniform thickness, shaped as in Fig. 2.26. It can be visualized as consisting of two rectangles with their respective centers of gravity at points *G* and *H*. The larger rectangle is twice the size of the smaller.

Suppose we suspend the body on a string attached at point *H* so that the

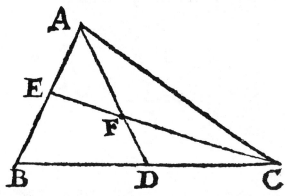

Figure 2.25 Two intersecting medians give the location of the center of gravity *F*.

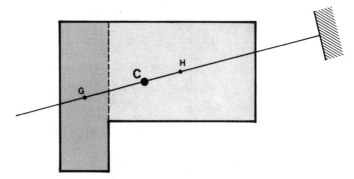

Figure 2.26 A composite body hanging from a string. The center of gravity *C* must lie on the line *GH*.

larger rectangle is completely supported by the string. The body will come to rest when point *G* hangs directly below point *H* (the string in the diagram should, of course, be vertical). Next consider the body as a whole: when it hangs from point *H*, the center of gravity *C* of the whole body must also rest on the vertical line passing through point *H*.

Where exactly on the line *HG* is the center located? The body is balanced around its center of gravity no matter how we turn it. If we suspend it from that point it will be in equilibrium just like a lever having the two rectangles hanging from its ends (Fig. 2.27). Because of the relative sizes of the rectan-

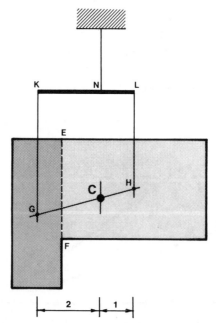

Figure 2.27 Back where we started from: imagine the composite body cut along line *EF*. The law of the lever gives the location of the center *C*.

gles the fulcrum must divide the imaginary lever in the ratio 2:1 as the law of the lever dictates. The center of gravity of the body must be vertically below the fulcrum. The similarity of our diagram to Fig. 2.3, which Stevin used to derive the law of the lever from an axiom concerning centers of gravity, is obvious. We have come full circle.

THE FORCE COUPLE

Consider a man in a boat near a dock. He wants to turn the boat and, to accomplish this, he pushes the boat away from the dock. In this attempt to turn the boat by means of a force the man's position in the boat is of importance: the rotation of the boat will take place in one direction if he stands in the stern and in the opposite direction if he stands in the bow. Either way, in addition to turning the boat, he will also move it in the direction of the push. If he stands directly above the center of gravity C of the boat, the man will only manage to move away from the dock and no rotation will take place.

There is another way of turning the boat. The man will turn it if he pushes against the dock with one hand and pulls with the other—in other words, if he exerts two equal and opposite forces along parallel lines of action (Fig. 2.28). Such a pair of forces has special properties and is therefore given a distinctive name: a *force couple*.

A glance at the diagram will convince us that it is now completely immaterial whether the man stands at the bow or at the stern. The same force

Figure 2.28 Turning a boat by pushing and pulling. No resultant force acts on the boat.

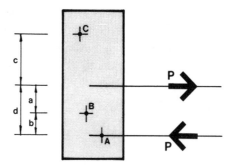

Figure 2.29 Special property of a couple: its turning effect is the same about points *A*, *B*, or *C*.

couple will always produce rotation in the same direction. Moreover, there will be no tendency for the boat to move away from or closer to the dock because the push and the pull (provided, of course, they are equal) will cancel out. Only the turning effect will remain.

Not only the direction but also the magnitude of turning effect of a couple—its *moment*—remain the same no matter where it is applied. This can be easily demonstrated by some simple calculations (Fig. 2.29). There are two equal and opposite parallel forces acting on the body in the diagram. Let us find their resultant moment about a number of points on the body. First, point *A:*

$$M_A = P \times d$$

Next, take point *B* between the two forces:

$$M_B = P \times a + P \times b$$
$$= P \times (a + b) = P \times d$$

same as before. Finally, let us take moments about point *C*, which is outside the couple:

$$M_C = P \times (d + c) - P \times c$$
$$= P \times d$$

The moment of a couple is equal to the product of the magnitude of the component forces and the distance between them for any center of rotation on (or off) the body.

The turning effect of a couple, its moment, does not change if the couple is moved to a new location on the body. We can also change the direction

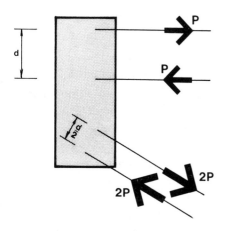

Figure 2.30 Where the couple is applied and how it is made up makes no difference.

and the magnitude of the two constituent forces of a couple and its effect will not change as long as the product

$$M = P \times d$$

and the sense of rotation (clockwise or counterclockwise) remain the same. For example, the two couples shown in Fig. 2.30 will have an identical turning effect on the body because their moments are the same.

Thus a couple is completely defined by the magnitude and sense of rotation of its moment. To represent a couple in a diagram all we have to draw is a curved arrow showing the sense of rotation with the magnitude of the moment marked beside it (Fig. 2.31). Where exactly on the body we draw our curved arrow makes no difference at all: the moment is always the same.

At first glance, the concept of the couple does not seem a discovery of major importance. Indeed, it will not help us solve any problems that we could not solve without it. What it does, however, is help us think clearly about the effect of forces.

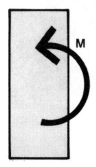

Figure 2.31 A curved arrow anywhere on the body can adequately represent a couple.

EXAMPLE

A typical tower crane (Fig. 2.32) has two jibs at the top of the tower: the saddle jib, carrying a movable saddle equipped with a hook for the load, and the counterweight jib. The weight of the saddle jib and the load on it tends to overturn the crane; the load at the end of the counterweight jib induces a moment of opposite sign in the structure. However, since the counterweight is a force of constant magnitude on a fixed moment arm while the load and its moment arm change, the two moments are usually different.

Two important consequences follow. First, the bottom end of the crane must be anchored to the ground by guy wires or bolted down. These supports provide the reactions required to oppose the resultant moment acting on the crane. Second, to minimize the resultant moment, the maximum load that the crane is allowed to lift is reduced with the operating radius of the load saddle: the farther out the saddle, the less load the crane is allowed to lift.

Figure 2.32   A tower crane. The moment of the load is opposed by the moment of the counterweight.

Example   41

The force of the wind, if it catches the jibs of a freestanding crane broadside, acts on a very long lever arm (the height of the crane) and could easily overturn or break the tall structure. To prevent this from happening, large name boards or similar items presenting a large area to the wind are restricted by crane manufacturers. Furthermore, during storms or whenever the crane is left unattended, the jib is allowed to turn freely in the wind like a giant windvane.

CHAPTER THREE _____

The Inclined Plane

"A mystery is not a mystery"

_____

THE LAW OF THE INCLINED PLANE

The contraption seen in Fig. 3.1 is a typical perpetual motion machine. The principle of operation is simple: a heavy chain is fitted around the wheels of the machine in such a way that the right part of the chain is always longer, and therefore heavier, than the left part. The inventor assumed that this imbalance would make the chain move and keep it going around indefinitely. Such inventions started to appear in the twelfth century and multiplied rapidly despite the fact that none of them ever worked, including the example here. The search for perpetual motion was always fruitless—but the profound insight that such motion is impossible led to valuable discoveries. One of them was the *law of the inclined plane*, discovered by Simon Stevin.

The problem facing Stevin was this: a body placed on an inclined plane (which is simply a ramp) somehow appears lighter than when it is sitting on a horizontal plane. For example, body *D* in Fig. 3.2 can be supported on the ramp *AB* by a considerably smaller body *E*. This makes the ramp a simple machine no less wonderful than the lever. Under ideal circumstances (no friction), a pull slightly greater than the weight of body *E* will move the load up the ramp. Thus with the help of the ramp we can raise a heavy load which, unaided, we could not lift.

How much force do we have to use to lift a given load up a ramp of known slope? While the law of the lever goes back to at least the days of Aristotle, the inclined plane defeated most writers before Stevin.

Stevin reasoned as follows: imagine a chain of equally spaced identical

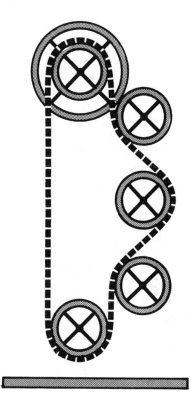

Figure 3.1 There is always more chain on the right side of this machine. Will the machine turn forever?

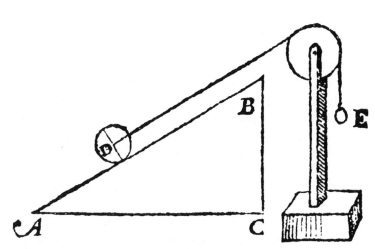

Figure 3.2 The paradox of the inclined plane: a small body *E* holds up a large body *D*.

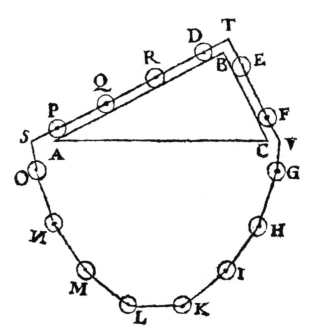

Figure 3.3 Perpetual motion is impossible therefore the chain of spheres will not move.

spheres draped over a triangle where side *AB* is twice the length of side *BC* (Fig. 3.3). The planes are without friction and the string is perfectly flexible.

We have here the prototype of the perpetual motion machine shown before. If the four spheres on side *AB* pull more than the two spheres on side *BC*, the chain will start to move and, since the situation does not change as each new sphere takes the place of the preceding one, it will have to go on forever. Our inventor hoped to make a fortune with his machine; other inventors were still trying in the nineteenth century. Stevin, in the sixteenth, starts by rejecting perpetual motion as absurd in principle. Nature just does not give something for nothing. The rest follows by pure logic.

If perpetual motion is impossible, the chain must be in equilibrium. We can cut off the bottom loop of the chain (spheres *G* to *O*), which is perfectly symmetrical and pulls with equal force at both ends. The two portions of the chain resting on *AB* and *BC*, even though of unequal length, must then balance each other.

Equilibrium will not be disturbed if we collect the spheres on the two planes and melt them into single but correspondingly larger spheres *D* and *E* (Fig. 3.4). Since there were originally four spheres on slope *AB* and two on *BC* the weight of the two aggregate spheres will also be in the ratio of 2 to 1. The two inclined planes that form the triangle can be at any angle and the proof would still hold. Make the side *BC* vertical so that the body *E* on that

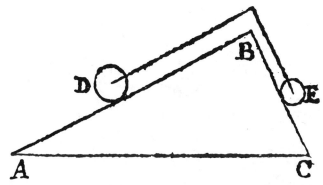

Figure 3.4 The spheres *D* and *E*, whose weight is proportional to the lengths of the sides *AB* and *BC* of the plane, will balance.

side hangs free (Fig. 3.5) and the rule still holds: if *AB* is twice the length of *BC*, the counterweight *E* need be only half of *D*.

In general, the number of spheres resting on either slope is proportional to the length of the slope and so, in this ingenious and highly original way, Stevin arrives at the law of the inclined plane:

$$E : D = BC : AB$$

The magnitude *E* of the force supporting a body on an inclined plane is to the weight of the body *D* as the height of the plane *BC* is to its length along the slope *AB*.

In Stevin's book, theory is always followed by a practical example. Suppose that the wagon in Fig. 3.6 weighs 10 kilonewtons. The slope of the road is such that the length of the inclined plane *AB* is four times its height *BC* (*BC* : *AB* = 1:4). The horse, pulling on a rope parallel to the slope, must exert a force of magnitude

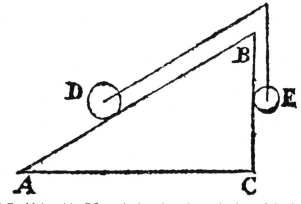

Figure 3.5 Make side *BC* vertical and we have the law of the inclined plane: *D* : *E* = *AB* : *BC*.

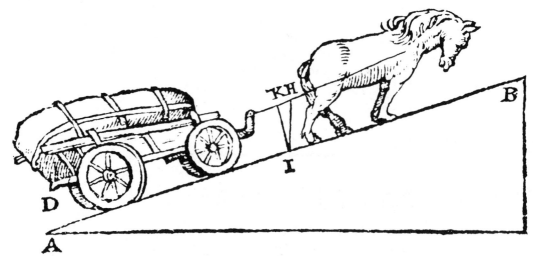

Figure 3.6 The law of the inclined plane applied to a practical problem. Note the triangle *KHI*, the first of many.

$$E = 10/4 = 2.5 \text{ kilonewtons}$$

to hold the wagon on the ramp. In the absence of friction, any small effort in addition to force **E** will pull the wagon up the slope.

Stevin was justly proud of his discovery. The chain of spheres appears on the title page of most of his books (Fig. 3.7) with an inscription that reads "a mystery is not a mystery." However, a significant development of the law of the inclined plane was hidden in the unobtrusive little triangle *KHI* squeezed in his illustration between the horse and the wagon.

THE TRIANGLE OF FORCES

The triangle first appeared in Stevin's diagram which is reproduced in Fig. 3.8. Stevin shows here a rectangular body supported on an inclined plane by a pull **E,** which is less than the weight **W** of the load as measured by the balance (but the balance does not support the body). Now look at the triangle *LDI,* formed by the two strings and the side of the body. Two of its sides (*LD* and *LI*) are at right angles to two sides of the triangle *ABC*, while the third side (*DI*) is parallel to the slope *AB*. It can be readily seen and also proved by geometry that triangle *ABC* and triangle *LDI* are similar. The sides of similar triangles have the same proportions regardless of size therefore if the weight *W* of the body on the plane is to the pull *E* as side *AB* is to side *BC*, then it is also as *LD* to *DI*:

$$W : E = AB : BC = LD : DI$$

SIMON STEVIN
van Brugghe.

WONDER EN IS GHEEN WONDER

Tot Leyden,
Inde Druckerye van Chriſtoffel Plantij
By Françoys van Raphelinghen.

Figure 3.7 "A mystery is not a mystery."

The useful thing about triangle *LDI* is that it has been turned relative to *ABC* through an angle so that the vertical force *W* is proportional to the vertical side of the triangle, while the force *E* is proportional to the side parallel to the direction of the string.

If we go back to Fig. 3.6 and the little triangle *KHI*, we notice that it is also similar to triangle *ABC*. Therefore, if the vertical side *HI* represents the weight

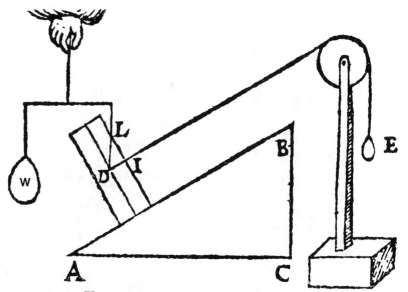

Figure 3.8   Triangles *ABC* and *LDI* are similar.

of the wagon (10 kilonewtons), then side *KH*, in line with the rope, will be proportional to the pull of the horse. Stevin's significant achievement here is that he has managed to separate the triangle *KHI*, which gives the ratio of the forces, from the inclined plane. This can be seen even better in Fig. 3.9 where only the slope of the plane is given: the ratio of the supporting pull *P* to the weight of the load is obtained from the triangle *OEC* whose two sides *OE* and *EC* are drawn parallel to the two forces and the third side (*OC*) at right angles to the plane.

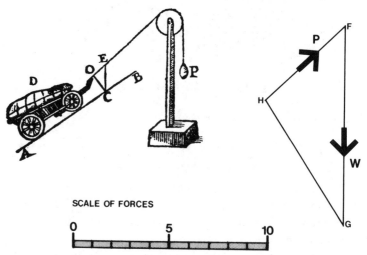

SCALE OF FORCES

0   5   10

Figure 3.9   Triangle *OEC* is also similar to the triangle of the inclined plane and so is triangle *FGH*.

For some reason—maybe economy—Stevin likes to insert these little triangles somewhere inside his diagrams. But since all similar triangles have
the same proportions regardless of size, let us draw a larger triangle *HFG*
outside, and draw it to scale. Then if the length of the vertical side *FG* is
made to represent to scale the known weight of the wagon *W,* the magnitude
P of the required pull in the rope can be scaled off the triangle. The arrows
in the sides of the triangle conveniently show the direction of the forces they
represent.

This is only the beginning. Next Stevin shows in a rather roundabout way
(which is somewhat modified here) that the third side of the triangle also
stands for a force. Figure 3.10 shows the forces acting on the prism sup-

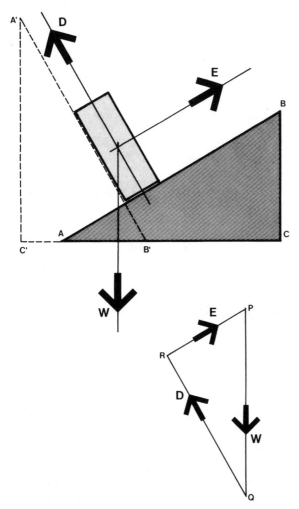

Figure 3.10 Proof that the third side of the triangle also represents a force. We
now have the "triangle of forces."

ported on the inclined plane without Stevin's accessories of Fig. 3.8. The weight of the prism and the pull of the string have the same ratio as the parallel sides PQ and RP of the triangle PQR which we add to replace Stevin's tiny triangle LDI. The third force acting on the prism is less obvious: it is the push of the inclined plane itself. The body is in contact with the plane and contact between two bodies always means a force. Since we assume the plane to be perfectly smooth, its push can only act at right angles to its surface. This force **D,** combined with the string force **E,** balances the pull of gravity on the prism.

Suppose now that the roles of the forces **D** and **E** are reversed: take away the inclined plane ABC and use the pull of a string instead to supply the same force **D** as before; instead of the weighted string let a new inclined plane $A'B'C'$ prop up the long side of the body with a force **E** (this new plane, shown by a dashed line, is really the original plane ABC stood on its side BC). The magnitudes of the forces **W** and **D** are proportional to the sides of the triangle PQR parallel to the forces:

$$W : D = PQ : QR$$

As we already know that side RP represents the magnitude of force **E,** we can simply state: the magnitudes of the three forces balancing a body on the inclined plane are proportional to the sides of a triangle parallel to their respective lines of action:

$$W : D : E = PQ : QR : RP$$

This triangle we call the *triangle of forces.*

Stevin's next step shows a remarkable depth of insight. He realizes that what he has discovered is far more general than the balance of weight and counterweight on a ramp (Fig. 3.11). The plane provides a push at right angles to its slope: this can be replaced by a string pulling in the same direction. The body on the plane can also be eliminated and a vertical force put in its place. The problem of Fig. 3.8 assumes a totally different appearance but, in essence, nothing has changed: the three forces acting at C must be proportional to the sides of the triangle HIC, each to the side parallel to its own line of action. We add a similar triangle outside.

The triangle of forces has thus been completely divorced from the inclined plane. Stevin discovered the *condition of equilibrium* for three forces meeting at a point: any three forces that can be drawn to form a closed triangle of forces are in equilibrium.

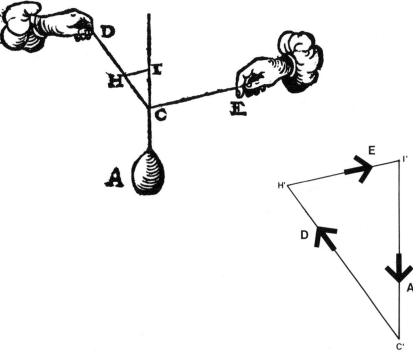

Figure 3.11   Stevin's ingenious application of the triangle of forces *HIC* and its more efficient modern version.

THE PARALLELOGRAM LAW

In the problem seen in Fig. 3.12 Stevin develops his ideas further. There is no reason why one of the forces should always be a weight force, directed vertically down. In this problem the three forces act in perfectly arbitrary directions (but a pulley is still needed to achieve that). Stevin is looking for the magnitudes of the pulls **D** and **E** required to support the load **A** whose weight is known. The solution is contained in the diagram.

Almost four centuries later, we use virtually the same approach. We draw a vertical line segment whose length represents, to a convenient scale, the magnitude of the force **A** which is known. We complete the triangle of forces by drawing lines parallel to the pulls **D** and **E** from the endpoints of force **A**. Two triangles can be obtained in this way; both are similar to triangle *HCI* in Stevin's solution and equally valid. Finally we scale off the magnitudes of the two forces we were looking for.

All seems perfectly logical and a suitable experiment will confirm the result obtained in this way. However, there is a gap between the law of the inclined plane and the triangle of forces derived from it when it comes to forces pulling in arbitrary directions. The triangle of forces is supposed to be

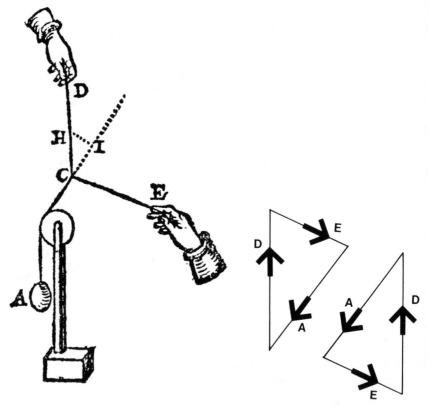

Figure 3.12   The triangle of forces in its most general form.

similar to the triangle representing the inclined plane—that is where it came from. But then it should always contain a right angle. The triangle in Fig. 3.12 does not; it corresponds to a body on an inclined plane held up by a string not parallel to the plane (Fig. 3.13). Stevin's proof is valid only as long as the string is parallel to the slope of the plane, as the famous chain of spheres was.

We crossed this gap between proof and application with no apparent support—and so did Stevin. He tried but never really managed to bridge it. Nevertheless, he applied the triangle of forces even when he had no real proof of its validity because, no doubt, he was intuitively sure that such application was justified. He was right, and the proof finally came—about a hundred years later. It was supplied by *Isaac Newton.*

Suppose that you are trying to row a boat across a river (Fig. 3.14). Rowing straight across you would manage in one minute to push the boat from *A* to *B* but, in the same time, the current would take it downstream from *A* to *C.* Obviously, with both forces acting on the boat you will end up at point *D.* "A body acted on by two forces will describe the diagonal of a parallelogram in

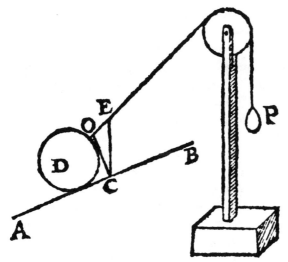

Figure 3.13 The stumbling block: this triangle of forces was never proved by Simon Stevin.

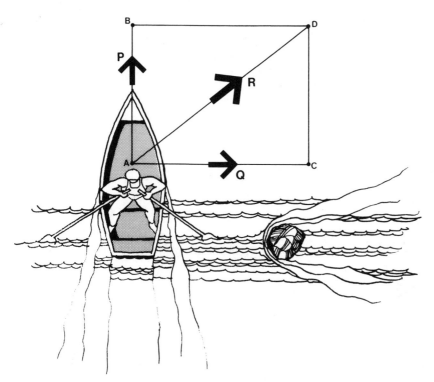

Figure 3.14 Newton's parallelogram law. Equilibrium or nonmotion is just a special case of motion.

the same time that it would describe the sides if acted on by those forces separately," stated Newton in 1687.

In the century between Stevin and Newton a crucial advance took place in mechanics: an understanding of the relation between force and motion. That was the beginning of the branch of mechanics called *dynamics*. Newton knew that forces are proportional to the motions produced by them in equal intervals of time. Therefore, if we make two sides of the parallelogram *ABCD* represent forces **P** and **Q** instead of motions produced by them, then the single force **R**, its magnitude and direction given by the diagonal *AD* of the parallelogram, completely replaces those two forces. This is Newton's principle of the *composition of forces*. Force **R** is called the *resultant* of forces **P** and **Q**, and forces **P** and **Q** are the *components* of force **R**.

Alternatively, the parallelogram construction can be reversed to *resolve* a single force into two component forces. For example, the body in Fig. 3.15 is suspended from a string which is pulled up by a single force **F**. Two strings along the lines *CD* and *CE* can do the same job. The forces in the strings can be found by resolving force **F** into components along those lines by completing the parallelogram *IHCK*. If the vertical pull is proportional to the diagonal *IC*, the forces in the strings will be proportional to the sides *CH* and *CK*. A larger parallelogram, drawn to scale rather than relying on proportion, should be used for better accuracy and convenience.

The diagram we just used came from Stevin's book but the reasoning was along Newton's lines. Stevin obviously knew about the parallelogram; he arrived at the construction by combining two triangles and pointed out that

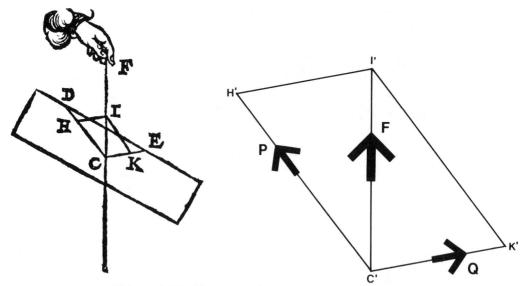

Figure 3.15   The components of a force. Two pulls, in the directions *CD* and *CE*, completely replace the vertical pull **F**. The parallelogram law gives their magnitudes.

one triangle was sufficient since side *HI* was the same length as side *CK* and side *IK* equal to side *HC*. But the parallelogram, as derived from dynamics by Newton, provides the logical basis for using the triangle: the triangle is only half of the parallelogram. On the other hand, Stevin would most probably have rejected Newton's line of reasoning in stating the parallelogram rule. What has motion to do with forces in equilibrium where no motion takes place? However, to his successors in the seventeenth century and to modern mechanics, equilibrium (nonmotion) is just a special case of motion.

THE ALGEBRA OF ARROWS

From the preceding discussion it is evident that, where the equilibrium of forces is concerned, we can do the same job using Stevin's triangle of forces or the parallelogram rule of Isaac Newton. They give the same results, only the reasoning behind the geometrical constructions differs. In modern mechanics, however, we use neither. Once again geometry is replaced by algebra, this time a branch of mathematics specially developed to deal with quantities that have direction as well as magnitude. Such quantities are called *vectors* and we are talking of *vector algebra*.

Vector algebra provides a set of rules for representing and manipulating all vectors (including forces) and a convenient mathematical shorthand for describing exactly what we are doing with them. It performs the same service for vectors that ordinary algebra does for nonvector (or *scalar*) quantities. The full benefits of this sophisticated tool can be appreciated only when we have to deal with forces in three-dimensional space, but here we make use only of the tip of the iceberg: vector addition.

Suppose that we want to find the combined effect of the two forces, **A** and **B**, seen in Fig. 3.16. In the language of vector algebra, we add vectors **A** and **B**. First, we draw vector **A** so that both the magnitude and the direction are contained in the diagram: we draw a line segment parallel to the line of action of the vector and give it a length equal to its magnitude, to scale (as we have been doing all along with forces). Next, to add vector **B** to vector **A** we place the tail of vector **B** at the tip of vector **A**. A new vector, joining the tail of **A** to the tip of **B**, is the sum or resultant **R** of the two original vectors. Using boldface letters to indicate that this is a vector operation, we write

$$\mathbf{A} + \mathbf{B} = \mathbf{R}$$

The same result is obtained if we place the tail of **A** at the tip of **B** and then complete the triangle with the resultant vector **R**. In vector, just as in scalar addition,

$$\mathbf{A} + \mathbf{B} = \mathbf{B} + \mathbf{A}$$

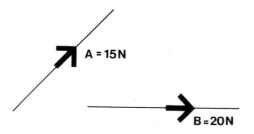

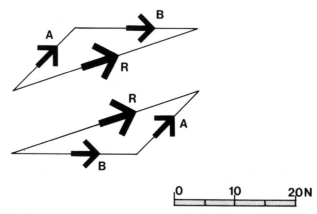

Figure 3.16 Two vectors and their sum: **A** + **B** = **A** + **B** = **R**. Note the direction of the sum vector **R**.

Obviously, vector addition represents no more than the parallelogram law separated into two triangles—if it did not, if it was different, we could not apply it to forces.

Note that in both cases of addition, while the component vectors **A** and **B** follow each other tip-to-tail, the resultant vector **R** is directed in the opposite sense: its tail starts at the tail of the first component vector and its tip joins the tip of the second.

Using vector addition, we can easily add any number of vectors. It simply means repeating the operation of vector addition on successive vectors, adding each new vector to the sum of the vectors added previously. In Fig. 3.17 a vertical force **C** (C = 20 newtons) is added to vectors **A** and **B**. We show the addition:

$$\mathbf{A} + \mathbf{B} + \mathbf{C} = \mathbf{Q}$$

where **Q** is the new resultant. The dashed line in Fig. 3.17 is the resultant **R** obtained before. It replaces the two components **A** and **B**, so that what we really have here is two additions:

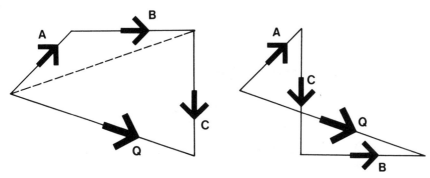

Figure 3.17   Vector addition: **A + B + C = A + C + B = Q.**

$$A + B = R$$

and

$$R + C = Q$$

in one diagram. The resultant **R** is not required so it can be left out and the addition of the three vectors performed directly by the basic general rule of vector addition, the *polygon rule:* to add any number of vectors, chain them together tip-to-tail in any order. The resultant vector, going in the opposite direction, closes the gap between the first and last vector to form a polygon. Figure 3.17 shows the same resultant **Q** obtained by the operation

$$A + C + B = 0$$

The triangle can now be seen as only a special case of vector addition: it is a polygon with just three sides.

When several forces are in equilibrium there is no net push or pull on the body on which they act. Their resultant force is zero and the body will not move, or *translate,* in any direction. If we form a polygon of vectors with our forces and find that they all go around tip-to-tail leaving no gap to be filled by a sum vector, then obviously the resultant is zero. Force **D** in Fig. 3.18 (being equal and opposite to **Q**) will have this effect and

$$B + A + C + D = 0$$

A general condition of equilibrium can now be stated: any number of forces which, when represented as vectors, can be drawn to form a closed polygon of vectors have no resultant force. The body on which they act is in *translational equilibrium:* it will not move in any direction. In the special case of

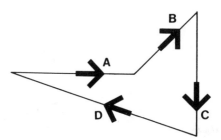

Figure 3.18 Vectors following each other tip-to-tail in a closed polygon. Their vector sum is zero: **A** + **B** + **C** + **D** = 0.

only three forces in equilibrium the polygon is a triangle with the three arrows chasing each other tip-to-tail. This is exactly what Stevin derived from the law of the inclined plane.

It is interesting to observe how much more we can do with vector algebra than we could with the triangle or parallelogram of forces alone—and yet no new discovery was involved. Like all mathematics, vector algebra is an ordering device and a means of succinct communication. It does not teach us anything about forces that we did not know before.

MORE SIMPLE MACHINES

A number of simple machines and other devices are based on the principle of the inclined plane. A very common application is the *wedge*. The normal ramp is usually fixed in position while the load is pulled up on it by an external force. In the case of the wedge the load is stationary while the wedge is pushed under it, thus raising the load. The smaller the slope, that is, the thinner the wedge, the larger is the force exerted against the load. A double wedge (Fig. 3.19) is commonly used as a cutting tool, forcing apart the sides of the opening in which it is inserted.

When we use a wedge to split wood, the force of friction that we neglected in the discussion of the inclined plane plays a very important part. The wedge is usually driven by blows rather than by a constant force. Without friction the wedge would bounce back after every blow. On the other hand, because of friction we would not have much success with our wedge without the dynamic effect of the blow: a very much larger force would be required to push it forward.

The wedge appears in many forms. Before the invention of the steam engine the two main means of propulsion of boats were the oar, which is really a lever, and the *sail*, which is another manifestation of the inclined plane or wedge. The principal difference is that the wedge pushes against the load and moves it, whereas in the case of the sail the opposite is true: the "load" (the force of the wind) is used to move the wedge.

Figure 3.19 The wedge is a double inclined plane.

Let us consider the forces at work in the process of sailing against the wind, one of the major inventions of the century that preceded the great voyages of discovery. Figure 3.20 shows in a highly simplified form the forces acting on a sailing boat. The force of the wind **W** may be replaced by two components: force **A** in line with the sail, which has no effect on the sail, and force **B** acting at right angles to the sail, which does all the pushing. To find the effect of the force **B** on the boat we have to resolve it into two components: a component **P**, which pushes the ship sideways, and **Q**, a component in the direction of the boat's forward motion. The boat is constructed with a deep keel giving it great resistance against sideslip but the resistance against forward motion is very small. Thus even though the component **Q** is comparatively small the boat will still move in the forward direction rather than sideways. As can be seen from the forces involved, we cannot sail directly against the wind but only close to it. To make headway, the ship has to follow a zigzag course.

A modified form of the inclined plane or wedge is the *screw*. Consider first a flight of stairs. This is also an inclined plane, although the ascent is produced by successive steps rather than a single smooth ramp. The steps are only a concession to the special way of lifting the load of the human body—by walking. Spiral stairs winding around a central column have the advantage that such an inclined plane does not occupy much space. The screw is simply an inclined plane wrapped around the surface of a cylinder.

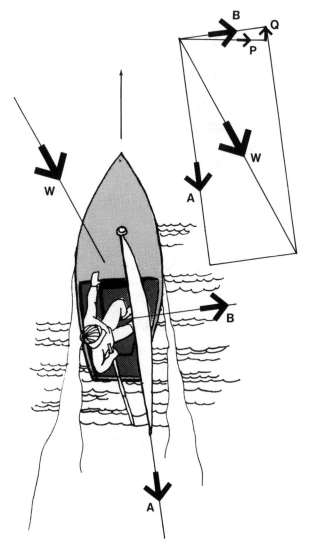

Figure 3.20   The sail is a sliding inclined plane.

There is a particularly striking similarity between the circular stairs and the *auger*, commonly used to drill holes for building foundations (Fig. 3.21). As the auger is screwed into the ground by the motor it raises a load of earth to the surface just like a continuous wedge.

Alternatively, the screw may turn inside a stationary housing. Then the screw itself moves forward with great force. An example is the *jack screw* (Fig. 3.22) used in various forms to lift heavy loads. It was evidently invented a long time before the automobile where we now find it as standard equipment. In this application, the screw is turned by means of a long handle so

Figure 3.21 An inclined plane wrapped around a cylinder can function as an auger and push up earth.

that the advantage of leverage is combined with the powerful action of the wedge—the jack is thus a *compound machine.*

When steamships first began to replace sailboats they were pushed forward by paddle wheels, which are just another application of the lever. A revolutionary development in ship propulsion was the invention of the *pro-*

Figure 3.22   A screw jack can push up all kinds of loads. If the direction of the push is reversed, the jack is converted into a press.

peller—a screw permanently forcing itself into the water and pushing the ship forward. The auger, drilling into the ground, works in exactly the same way—the fact that the propeller does not look like an auger is due to the circumstance that water is somewhat different from earth.

In other applications, the screw turns in place and its rotation is used to propel water forward. Such a device is the *Archimedean screw,* which consists of an augerlike screw enclosed in a tube (Fig. 3.23, at bottom). It pushes water up like the auger pushes up earth and has been in use for centuries. In the machinery shown in the illustration a water wheel is used to operate a series of Archimedean screws which raise water to a considerable height. Many a hopeful inventor tried to achieve perpetual motion by using the energy of the water thus raised to drive the water wheel.

Figure 3.23 Archimedean screws used to raise water. A cutaway view of one in the foreground—an auger enclosed in a tube.

PRINCIPLES AND LIMITATIONS

At first glance, the inclined plane seems to be a very different type of problem from the lever, involving totally different principles. Consider, however, Fig. 3.24, taken from a book on mechanics by the French mathematician Pierre Varignon (1687). Force **Q** is a known load on the lever and the problem is to find the magnitude **P** of the force holding the load up. The principle of moment equilibrium will give us the answer once we know the moment arms of the forces, but Varignon did the same job in a different way.

The effect of the two forces is not changed if they are both replaced by their resultant **R** which must pass through point A where the two component forces, **P** and **Q**, meet. If **P** and **Q** do not cause rotation of the lever, neither does their resultant. But the lever will not rotate only if the resultant **R** is directed at the fulcrum *f* of the lever so that its moment arm is zero. Given the direction of the resultant (along line *AC*) and the magnitude of **Q** (represented by the side *AD*) Varignon completed the parallelogram of forces *ABCD*. The length of the side *AB* represents the magnitude of force **P**.

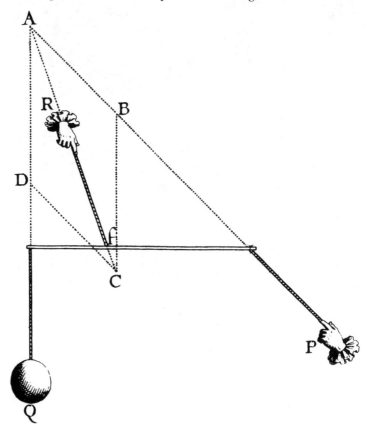

Figure 3.24   Varignon finds the forces acting on a lever by means of the parallelogram law. The resultant force must push on the fulcrum or the lever will rotate.

Newton and Varignon (who arrived at the principle of the parallelogram of forces at the same time) built all of statics on the parallelogram law. Since Stevin had previously derived the triangle of forces and the parallelogram from the inclined plane, it can be said that the whole subject is just a series of consequences following from the fundamental axiom that perpetual motion is impossible.

On the other hand, Galileo managed to explain the inclined plane by means of the lever by pointing out that the magnitude of the force **E** holding a body of weight **D** in equilibrium (Fig. 3.25) depends only on the slope of the spot on which the body is actually sitting. The rest of the plane is irrelevant. The same kind of support would be provided by a bar at right angles to the plane (bar *ST*). Where the plane pushes on the body with a force **N**, the bar would pull with the same force and completely replace the plane.

The magnitude of force **E** required to balance force **D** can now be found from the condition of moment equilibrium. Bar *ST* is the moment arm of force **E**, while force **D** acts on the moment arm *SV*. Then

$$E \times ST = D \times SV$$

since the two forces have a zero resultant moment about point *S*. Note that the triangles *ABC* (the inclined plane) and *SVT* are similar, therefore it is also

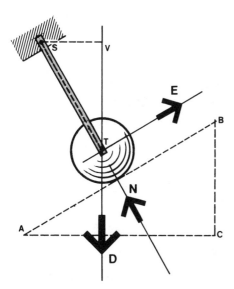

Figure 3.25 Galileo derives the law of the inclined plane from the principle of moment equilibrium. The rod *ST* completely replaces the inclined plane.

true that

$$E \times AB = D \times BC$$

and

$$E : D \ = BC : AB$$

which, of course, is the law of the inclined plane.

Thus the law of the inclined plane—and consequently the triangle of forces and vector addition—can be derived from the law of the lever. Hence all statics can be derived from the axiom that a uniform prism is balanced when supported at its center of gravity.

Every problem in statics can be attacked from different angles. The angle of attack chosen depends on the nature of the problem. The answer will always be the same; only the work involved in reaching it may be lessened by the most suitable approach. Some problems, however, remain unsolvable. For example, the load **A** supported by the three strings in Fig. 3.26 is in equilibrium, yet Stevin shows that there is no way we can determine the magnitude of the three forces in the strings required to balance it.

This does not mean that we cannot find a solution that would satisfy the conditions of equilibrium. On the contrary. Suppose force **F** is eliminated. The load will still hang in the same place because forces **D** and **E** are perfectly capable of holding it up, even though their magnitudes will be increased. But force **F** can assume an infinite number of different magnitudes to assist the other two forces and so the problem has an infinite

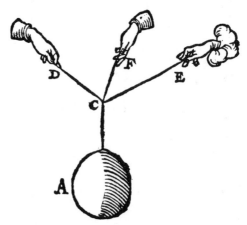

Figure 3.26 Some problems cannot be solved by statics alone. The string force **F** is statically indeterminate.

Example   67

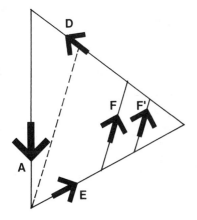

Figure 3.27 Force **F** can assume any magnitude from zero to a maximum shown by the dashed line.

number of possible solutions. There is just no way of deciding (by statics alone) which is the right one.

The same conclusion is reached if we try to use the vector polygon (Fig. 3.27). Just two possible alternative magnitudes of force **F** are shown in the diagram, both of them satisfying the condition that the resultant of the forces be zero. As the magnitude F becomes progressively smaller, the other two (D and E) become larger. Finally, when force **F** shrinks to nothing, the polygon of forces becomes a triangle involving only forces **A, D,** and **E.**

The same situation occurs whenever a body is held up by more than the minimum number of supports. Such problems are called *statically indeterminate.*

EXAMPLE

The ordinary domestic faucet works on the same principle as a jack screw. To stop the flow of water we turn the handle of the faucet, which is connected to a screw. The push of the screw forces a compressible washer against the opening of the supply pipe. This shuts off the water.

Figure 3.28 shows a so-called *gate valve,* often used in industrial installations where fluid pressures are great and where changing a washer would be a problem. The valve is made entirely of metal and contains no washers. It operates as a *compound* simple machine. The valve makes use of a screw turned by a large-diameter wheel to produce a very powerful force in the direction of the spindle of the valve. The large wheel, of course, adds leverage to the turning force applied to the screw.

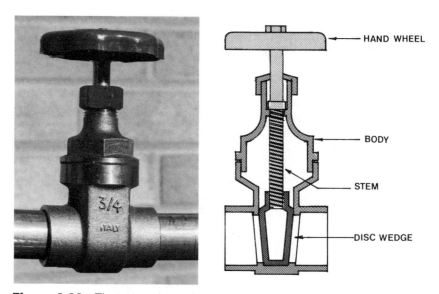

Figure 3.28 The gate valve makes use of both a wedge and the screw. The wedge is hollow and climbs up on the stem when the wheel is turned to open the valve.

The flow of the fluid is stopped by means of a retractable wedge, which is machined to seal tightly against two sloping metal surfaces. The large vertical force produced by the screw is transmitted to the wedge and jams it against the openings.

CHAPTER FOUR

The Beam

And a general method of attack

THE SUPPORT REACTIONS

Stevin's rectangular beam of 6 newtons, already familiar from Chapter 2, is shown once again in Fig. 4.1. This time it is supported by two strings. Figure 4.2 shows the same beam resting on two pyramid supports.

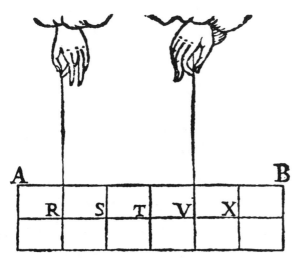

Figure 4.1   The strings must pull up with just the right forces to hold up the beam.

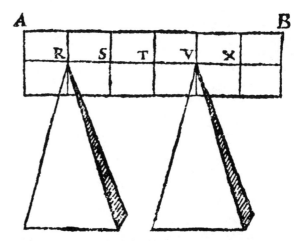

Figure 4.2 The beam is in equilibrium, therefore the pyramids must be pushing up with just the right forces—but how do they know what forces to exert?

The problem that occupies Stevin here is different from those he tried to solve in connection with simple machines. He is not after multiplication of force; this time he wants to find out how much of the load is supported by each string. To him, the problem is just another application of the law of the lever and, indeed, he manages to find the answer without too much trouble. The principle of moment equilibrium gives the same result more quickly. The pull of the string at point V (force $\mathbf{Q}$) has a counterclockwise moment about point R, opposing the clockwise moment of the weight force acting at the center of gravity T. Therefore, taking moments about point R,

$$6 \times 2 - \mathbf{Q} \times 3 = 0$$

and

$$\mathbf{Q} = 4 \text{ newtons}$$

We could now consider the body supported by the string at point V as on a fulcrum and find the force in the string at point R by taking moments about point V. Stevin takes a different route leading to the same result. If the string at V carries 4 newtons and the body is pulled down by a total weight force of 6 newtons then, reasons Stevin, the string attached at point R evidently carries the remaining 2 newtons. The problem is solved.

What is of interest now is that Stevin assumes that once he knows the forces in the strings, he also knows just how much force is exerted on the body by the pyramid supports. He takes it for granted that the pyramids, holding up the body at the same points as the strings, push up with as much force. This also seems a very acceptable assumption, but let us examine the implications. When a person pulls up on the strings by hand, he consciously adjusts the forces in the strings until the beam comes to rest. The forces have to be just right: a bit more or less force in one string and the beam will go up or down on that side. Furthermore, if we substitute a heavier beam, one weighing maybe 12 newtons, the pull of the hands doubles. Yet when the heavier beam is placed on the supports it remains in equilibrium, showing that the inanimate pyramids perform the same adjustments as the human arms. *How do the pyramids know what force to exert?*

There is no answer to this question and no explanation. It is simply a fact that they do. Stevin accepted it intuitively and we do the same in everyday life. It took the genius of Newton (Fig. 4.3) to recognize this fact as a fundamental physical law and formulate it in his so-called *law of action and reaction:* two bodies in contact push or pull each other with equal and opposite forces. The beam is pulled up by the string exerting a certain force on the body; the hand at the other end of the string feels an equal force pulling down. Similarly, the beam pushes down on the pyramid support with some force; the support responds with a reaction of equal magnitude acting in the opposite direction. Whenever the force exerted by the body on the support changes, automatically the support reaction changes too.

These faithful supports will keep pushing up with just the right force needed to cancel the downward push of the beam. They keep the beam in equilibrium. But this is only half the picture. The forces exerted by the beam on the supports form the other half. The pyramid on the left is being compressed by a force of 2 newtons, the one on the right with 4 newtons. When we know these forces we can design or select supports capable of resisting them. The abstract rectangle on pyramids drawn four hundred years ago might represent part of a building, a beam on two supports (Fig. 4.4). The forces on the supports would be found in the way just shown.

Action and reaction are not only equal and opposite but also distinct. The action of the beam on the posts depends only on the weight of the beam and possible loads on it and is not in the least affected by the supports. They may be made of steel or of cardboard: the action of the beam on them will be the same. The reaction, on the other hand, depends entirely on the support that is supplying it. It must always be equal and opposite the action—as long as it is there at all. If the action becomes excessive, the support will collapse; the reaction will no longer match the action and, inevitably, the state of equilibrium will end. The structure will fail.

Structures also fail in another way: by internal fracture. Stevin and Galileo extended the investigation of the equilibrium of the lever and the inclined

Figure 4.3 Isaac Newton (1642–1727). From a portrait painted in 1725 when Newton was 83.

plane to include forces acting not only on structures but also *inside* them. The method they employed, fully developed, is the cornerstone of modern mechanics.

THE THREE-FORCE BODY

Beams may be supported in many different ways. The beam in Fig. 4.5 is again suspended from strings but, in this case, the two men who supply the

Figure 4.4 Precast concrete beam being lowered in position. Supports must be designed strong enough to provide the required reactions.

string forces pull obliquely, in the direction of the lines *GK* and *HL*. Can we again apply the law of the lever or the principle of moments to find the support forces? Not quite. There is a difference.

All the simple machines we have seen so far had to move in predetermined ways because they were in contact with other bodies. For example, the lever was forced to rotate about the fulcrum, and the body on the inclined plane had no option but to follow the slope of the plane. However, the beam hanging from strings is perfectly free to move in any direction under the action of the string forces—it is a *free body*. When only three forces act on a free body (the force of gravity and the two string forces in this case), it is called a *three-force body*.

Stevin finds that it is not possible to hold up a free body by pulling on the strings in arbitrary directions. The directions of the pulls seem somehow linked together: when one of the strings is vertical, the other one must be vertical too. If the angle of one string changes and becomes oblique, the

Figure 4.5   The directions of the pulls holding up the body are linked. Change one and the other must change too.

other string must change direction too. But what direction will it take? In a typically ingenious argument Stevin proves that the body will come to rest only when a special condition is satisfied by the oblique forces (let us call them **P** and **Q**): their lines of action must meet at a point lying on a vertical line passing through the center of gravity of the body (Fig. 4.6).

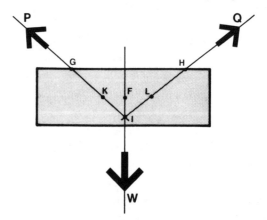

Figure 4.6   Stevin finds that the two supporting forces must meet on the line of action of the weight force.

We know that the line of action of a force can be extended and the force applied at any point on it without change in its effect. For example, force **Q** pulling at point *H* can be applied to a long or short string or the string can even be attached at point *L* and the effect of the pull will remain exactly the same as long as its direction and magnitude are not changed.

Let us imagine then that the two strings are extended until they are both attached at point *I*, where they meet. Their effect remains the same but the beam is now supported at this one point only. For the body to be in equilibrium supported at a single point, this point must be either directly above or directly below the center of gravity *F* or the body will overturn. In other words, the two string forces must meet on a vertical line passing through the center of gravity *F*. This is precisely what Stevin set out to prove.

Following this rule, if we change the direction of force **P**, the direction of force **Q** must also change to preserve equilibrium (and vice versa). Figure 4.7 shows the angles of the strings supporting the beam reversed. Still, the three forces acting on the body meet at a point on the vertical line through the center of gravity of the beam, as shown. This vertical line is, of course, the line of action of the weight force, so we can also state a general rule: if only three forces act on a body in equilibrium, their lines of action must meet at a point.

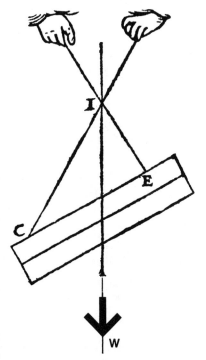

Figure 4.7 The forces meet at a point regardless of how the strings are arranged.

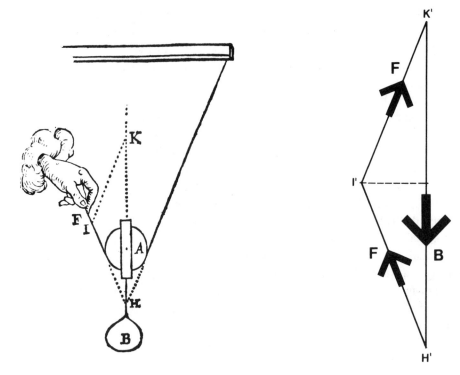

Figure 4.8 Once the directions of the forces are known, the vector triangle gives their magnitudes.

When the directions of the three forces on the body are known and the magnitude of the weight force is given we have all the information we need to find the magnitudes of the support reactions by vector triangle. Stevin does this in the following example. The pulley in Fig. 4.8 is held up by a force **F** in the string which is the same in the entire string. Since the force on one side of the pulley is not vertical, the string on the other side must also assume a slanting position so that the two equal string forces meet at point *H*, on the line of action of the load **B.** The triangle of forces *IKH*, inscribed in the diagram, gives the relative magnitudes of the three forces.

Let us assume that the magnitude *B* of the load is 100 newtons and use vector addition in a separate diagram to find the string forces directly. The weight force is drawn to scale and the magnitudes of the string forces measured off the vector triangle, which is, of course, similar to triangle *IKH*. Note that since the force in the string is the same on both sides of the pulley (it is the same string and there is no friction), the slope of the two ends of the string must also be the same because the sides in the vector triangle representing the pulls are of the same length.

Now that we have established, with Stevin, the rule of equilibrium for three-force bodies, another question springs to mind: why, in the preceding section, did we treat the beam hanging from vertical strings differently? After all, that beam is also a three-force body. Should the three forces acting on it not meeet at a point? In the next section we discuss the conditions of equilibrium that apply to all free bodies, regardless of the number and direction of the forces acting on them.

THE CONDITIONS OF EQUILIBRIUM

Let us examine Stevin's argument concerning the equilibrium of the free body. It is a clever argument, but there is really no need for a *new* line of reasoning.

The study of the problem of the inclined plane revealed that when a number of forces act on a body there will be no net push or pull (no resultant force) when, in the language of vector algebra, the vector sum of the forces acting on the body is zero. Now suppose the two men pull as seen in Fig. 4.9. The directions of their pulls are arbitrary and they meet at point N, not on the weight force. Still, the men can adjust their pulls so that the three forces acting on the body have a vector sum

$$\mathbf{P} + \mathbf{Q} + \mathbf{W} = \mathbf{0}$$

There is no net resultant force pulling or pushing on the body.

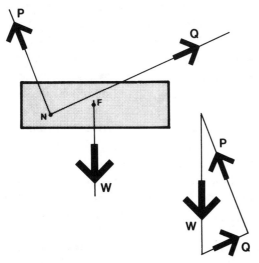

Figure 4.9 The resultant force is zero but there is nothing to balance the moment of force **W** about point **N**.

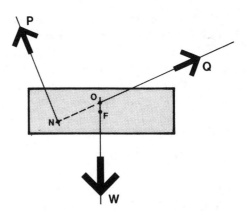

Figure 4.10   There is nothing to balance the moment of force **P** about point O, or the moment of force **Q** about the point where forces **P** and **W** meet.

But is the body in equilibrium? The fact that there is no net force on the body means that the body will not be moved up, down, or sideways. This is called *translational equilibrium*. But a body can remain in place and rotate.

Questions of *rotational equilibrium* take us back to the old law of the lever and to its modern expression in terms of moments of forces. A body does not rotate when the resultant moment of the forces acting on it is equal to zero. The three-force body can fulfill this condition only when all the forces acting on it pass through a single point because there is no other way they could have a zero resultant moment. If one of the forces did not pass through the point where the remaining two meet, it would have a turning moment about that point and cause rotation. Thus force **W** in Fig. 4.9 would cause clockwise rotation because it has a moment about point N. Looking at it another way, force **P** (Fig. 4.10) would cause clockwise rotation because it has a moment about point O where the remaining two forces meet.

At first glance, it may seem strange that the three forces should cause rotation even though their resultant is zero. Let the two string forces be replaced by their resultant **R** (Fig. 4.11). The two equal and opposite forces acting on the body (the resultant of the string pulls and the weight W) form a *couple*. It has no resultant force but only a resultant moment: $M = R \times d$ or $M = W \times d$, which amounts to the same thing because the two forces, **R** and **W**, have equal magnitudes. The effect is the same as if somebody tried to raise the body by pulling up at point N, away from its center of gravity. Equilibrium is impossible even though the body will not fall down: it will turn.

What we have just found applies to any number of forces. If the resultant force of a system of forces acting on a body is zero, the body will not translate: its center of gravity will remain in the same place. The body may

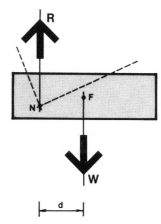

Figure 4.11   The resultant of the supporting forces and the weight force form a couple.

still rotate if the forces combine in a couple which has no resultant force but only a resultant moment. When the three forces intersect at point *I* they have no resultant moment about that point, and this also means that there can be no couple acting on the body. We know from Chapter 2 that the moment of a couple is the same for all points. If it is zero for any one point (point *I* in our example), it is zero everywhere.

When more than three forces act on a body they need not meet at a point for their resultant moment to be zero. This requirement applies only in the special case of a three-force body. The general condition of rotational equilibrium is simply that the resultant moment of the forces be equal to zero for any point. This guarantees that there is no couple acting on the body and that it is in rotational equilibrium.

Consider, for example, the body in Fig. 4.12. There is a weight force of 6 newtons acting at point *T* and, together with the loads hanging from it, this is balanced about the fulcrum at point *X*. The resultant moment about the fulcrum is zero:

$$- 6 \times 2 - 1 \times 1 + 13 \times 1 = 0$$

However, the body is in equilibrium not only because the moments of the forces balance about the fulcrum. This is only part of the picture. The upward force at the fulcrum *X* must also balance the total pull downward so that the resultant force is zero: the hand must pull up with a force of 20 newtons. If we take into account *all* the forces acting on the lever, including the upward force at the fulcrum, we find that their resultant moment is zero not only about the fulcrum but also about any other point. Thus the resul-

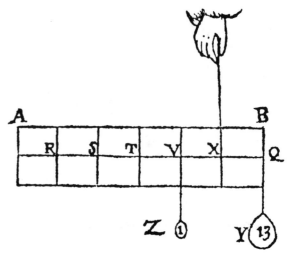

Figure 4.12   The sum of the moments of the forces acting on a body in equilibrium is zero about any point. The weight force of 6 newtons pulling down at point **T** is not shown.

tant moment about point *R* is equal to

$$6 \times 2 + 1 \times 3 - 20 \times 4 + 13 \times 5 = 0$$

A quick check will reveal that the resultant moment is zero about any other point too.

We can now state very concisely the conditions of equilibrium that apply to a free body with any number of forces acting on it. There are only two. *First,* the vector sum of the forces must be zero; if this condition is fulfilled, the resultant force will also be zero and the body will not translate. For three forces, this means that they have to form a closed vector triangle. If there are more than three forces acting on the body, they must form a closed polygon. *Second,* the sum of all moments about any point must be zero. This condition guarantees that the forces acting on the body cannot be reduced to a couple and the body will not rotate. In mathematical shorthand:

$$\Sigma \mathbf{F} = 0$$
$$\Sigma M = 0$$

The sign Σ (the Greek letter sigma) stands for summation.

Sometimes, instead of performing vector addition, we shall find it more convenient to resolve all the forces acting on a body into their vertical and horizontal components. Then the first condition of equilibrium takes a slightly different form:

$$\Sigma H = 0$$
$$\Sigma V = 0$$

Obviously, if there is no tendency for the body to move in the horizontal or in the vertical direction, the body will remain at rest.

The answer to the question posed at the end of the preceding section is now evident. The horizontal beam with vertical forces acting on it (Fig. 4.1) is subject to the same conditions of equilibrium as any other body. The sum of the moments about any point (we took moments about point R) must be zero. The sum of vertical forces (since there are no other) must be zero. The rule for a three-force body holds in this case too, even though the forces acting on the beam are parallel. In geometry, parallel lines are considered to meet at a point, only this point is infinitely far.

THE FREE-BODY DIAGRAM

The laws of the lever and of the inclined plane were immediately useful to their discoverers because they could be directly applied to simple machines, vitally important in lifting and moving heavy loads. The usefulness of knowing the conditions that govern the equilibrium of a free body is less evident, yet once we make full use of the concept of free body *we no longer need any of the old laws.*

The free body is the key to modern mechanics. In nature, there is a wide variety of machines and structures, created either by nature or by man. We would not get very far if we tried to analyze each of them individually. In mechanics everything is reduced to just *free bodies* and *forces.* This amazing feat of simplification is performed by means of a very efficient tool called the *free-body diagram*—an invention equal in importance to the discovery of the principles but not directly attributable to any single individual. We can best explain it through an example which also shows that Stevin intuitively understood the idea.

In the problem seen in Fig. 4.13 Stevin has a body supported by means of a pyramid at point E and a pull **H** at point F. He knows the weight of the body and wants to find out the magnitude of the force **H** that will hold the body in equilibrium and also what the support reaction is at point E. Let us solve the problem with the help of a free-body diagram. The procedure is always the same, as follows.

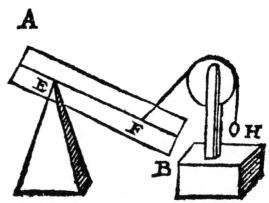

Figure 4.13   What are the support reactions holding up the beam?

Step 1.   We mentally isolate the body we wish to analyze from all other bodies it is in contact with and also from the influence of the earth. Thus we obtain a free body. Figure 4.14 shows the beam of the previous figure isolated as a free body. It is detached from its support and from the string and is free to move in any direction.

Step 2.   In Fig. 4.15 all those bodies removed in Step 1 are replaced by the forces they exert on the free body. First we replace the earth by the force it exerts on the body: the weight force **W** pulling vertically down at the center of gravity. This force is fully known, since we know both its magnitude and its direction. Next, the string is replaced by the force **H,** pulling in the direction of the string at point *F.* We know the direction of this force, but not its magnitude, which we mark by *H.*

So far, this has been fairly obvious. The important thing to remember now is that every body exerts a force on every other body it is in contact with. The pyramid support is in contact with the body we are analyzing so there must be a force at point *E*—the support reaction **G.** This is a completely unknown force. We know neither its direction nor its magnitude but we know that it must be there and we show it as an arrow with a squiggle on its shaft (meaning that the direction is purely a guess) and a letter (*G*) as a symbol of its magnitude.

This completes the free-body diagram. It shows very plainly that this is a three-force body.

Step 3.   The body is in equilibrium, therefore the conditions of equilibrium are satisfied. This fact yields the answers to the questions posed by Stevin (Fig. 4.16). There is no resultant moment on the body ($\Sigma M = 0$),

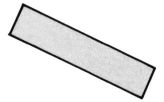

Figure 4.14 The beam as a free body. All contact with other bodies has been eliminated.

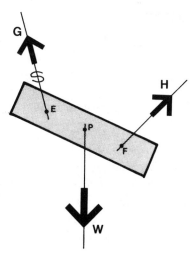

Figure 4.15 The free-body diagram of the beam. All bodies in contact with the beam are replaced by forces.

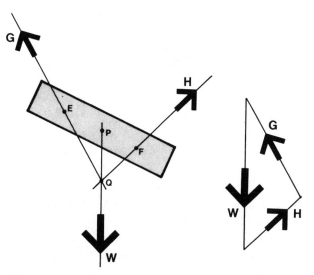

Figure 4.16 The conditions of equilibrium applied to the free body give us the required support reactions.

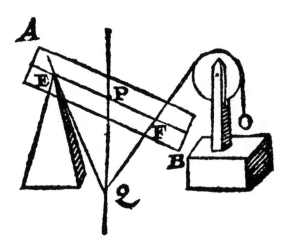

Figure 4.17   Take away the props and Stevin's solution is identical to the modern one.

therefore the three forces must meet at a point. Forces **H** and **W** meet at point Q, therefore force **G** also passes through that point. The direction we assumed for it must be corrected. There is no resultant force on the body ($\Sigma\mathbf{F} = 0$). The three forces therefore form a vector triangle which can be constructed now that the direction of force **G** is known. If the known weight force (6 newtons) is drawn as a vector we can scale off the magnitude of the unknown support reactions.

Figure 4.17 presents Stevin's solution, virtually identical to the modern analysis in the preceding diagram. Stevin arrived at this result by deep insight and logical reasoning. All we did was to follow a set of rules, step by step. No effort of imagination and very little thought was required. Moreover, all problems in statics are attacked in this same way.

Does Stevin's little beam-and-pyramid problem seem far removed from practical needs? Let us look at the saddle jib of the crane in Fig. 2.32. What are the forces in the steel cable and the jib bearing when the crane is lifting a load **W** at a given distance from the support? Isolate the jib as a free body (Fig. 4.18). Neglecting the self-weight of the jib and saddle, we find that there are just three forces acting on the body. Two of these, the pull **T** of the cable and the weight **W** of the load, intersect at point N. The force **R** acting on the bearing A must pass through the same point. The vector triangle gives us the magnitudes of the support forces right away. The support reaction **R** is further resolved into its components, showing the large horizontal component $\mathbf{R_h}$ that has to be resisted by the jib bearing.

The philosopher Alfred North Whitehead once wrote: "It is a profoundly erroneous truism, repeated by all copybooks and by eminent people when

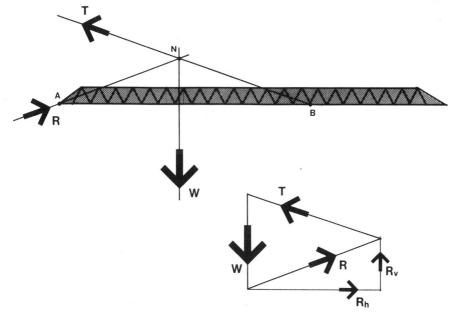

Figure 4.18 The forces acting on the jib of a crane. The support reactions change as the load moves along the jib.

they are making speeches, that we should cultivate the habit of thinking of what we are doing. The precise opposite is true. Civilization advances by extending the number of important operations which we can perform without thinking about them." Civilization took a giant step with the invention of the free-body diagram.

GALILEO'S PROBLEM

A famous illustration from Galileo's *Two New Sciences* (Fig. 4.19) shows a beam firmly embedded in a vertical wall at one end and supporting a load at its other, free end. Such a beam is called a *cantilever*. If the cantilever breaks under the load, it is clear (says Galileo) that fracture will occur across section *AB*. The beam acts as a lever with its fulcrum at *B*; the length *BC* is the lever arm of the load **E** while the depth *AB* of the beam is the other arm of the lever along which is located the resistance to fracture. The lever is shown in Fig. 4.20. The resistance is a force, labeled **R,** that opposes the pulling apart of the beam. Galileo assumes that it acts in the middle of the depth *AB* of the beam.

Using the law of the lever Galileo obtains the relation between load and resistance:

$$E \times BC = R \times 1/2\ AB$$

Figure 4.19 A cantilever beam from Galileo's *Two New Sciences.*

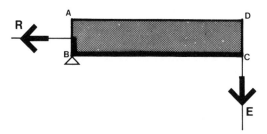

Figure 4.20 The lever inside the beam, according to Galileo. He assumes a fulcrum at point *B*.

This explains something we know from experience: a beam is easier to break in bending than in tension. The bending load **E** acts on the long arm of the lever (*BC*) while the arm of the resistance **R** is very short (½ *AB*). The shallower the beam, the easier it breaks in bending. For simplicity the weight of the beam itself is initially left out of Galileo's discussion.

Let us examine what Galileo is doing. By means of an imaginary break he cuts his beam into two parts. The part outside the wall, to the right of line *AB*, is treated as a free body. The part inside the wall, to the left of line *AB*, serves as a support for the free body just as if the two parts were glued together. The resistance **R** is treated as the support reaction which automatically adjusts to the load on the body and balances it. Galileo is doing exactly the same thing as Stevin in finding support reactions, only on a higher level of abstraction: he is not restricting himself to actual supports, visibly distinct from the supported body. Any part of a body supports all the adjoining parts; what holds them together are *internal reactions*. By knowing what these internal forces must be an engineer can design a structure strong enough to oppose the loads.

Galileo's pioneering contribution to engineering mechanics was this brilliant development and refinement of *method,* an anticipation of the full possibilities of what we now call the free-body diagram. His analysis of the cantilever, however, was inaccurate because in his mind rotational equilibrium was still intimately linked with the physical lever. We saw how Stevin, starting with the inclined plane, went far beyond to develop the triangle of forces as a generally applicable tool. For some reason, the much earlier law of the lever did not receive such treatment. The turning effect of a force, its moment, was not divorced from the simple machine.

Let us use Galileo's idea of an imaginary break, or cut, along the surface *AB* in the cantilever and isolate the block *ABCD* as a free body (Fig. 4.21). The isolation must be complete: there must be no contact with other bodies such as the fulcrum at *B* which Galileo needed because he was looking for a lever. The load **E** at the free end is shown as a force. The weight of the beam is left out to simplify the analysis. What other forces are acting on the body? It is in contact with the rest of the beam all across the surface *AB*,

Figure 4.21   The free-body diagram of the portion *ABCD* of the beam. Stress forces act at every point of the contact surface *AB*. Self-weight of the beam is neglected.

and at every point of contact there must be a force pushing or pulling on the free body. These forces are called *stress*. We know nothing about them except that they must exist and therefore we show them acting at random. This completes the free-body diagram.

The free body is in equilibrium under the action of those forces, therefore the conditions of equilibrium are satisfied. Now even though the stress forces themselves remain unknown, we can draw conclusions about their resultants. We shall look at the effect of their vertical and horizontal components separately. There is no net resultant force on the body, therefore the vertical components of the stress forces across *AB* must have a resultant **V** equal and opposite to force **E** (Fig. 4.22). This is called the *shearing force*. The horizontal components of the stress forces are still unknown and shown acting at random.

The shearing force alone cannot keep the free body in equilibrium. The two vertical forces (the load and the shearing force) form a couple, which on its own would cause the free body to rotate clockwise. There is no resultant moment acting on the body, so the inevitable conclusion is that there must also be a counterclockwise couple acting on the body (Fig. 4.23). The horizontal components of the stress forces add up to a pull somewhere at the top of the section *AB* (causing tension there) and an equal push at the

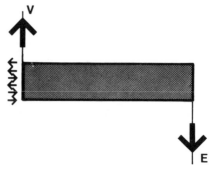

Figure 4.22   Vertical components of stress add up to a vertical shearing force in the section.

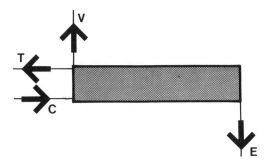

Figure 4.23 Horizontal components of stress form a couple that opposes the rotation of the free body. This causes tension in the top fibers of the beam and compression in the bottom fibers.

bottom (causing compression). The typical cross section of a steel beam reflects the resisting forces: it consists of two flanges located as far apart as possible, thus providing a large internal moment arm for the tension–compression couple, connected by a vertical web that takes care of the shearing force.

The two couples, the action and the reaction, cause bending and therefore are called the bending moment. Where exactly the forces **T** and **C** act we cannot say; that problem goes beyond the limitations of statics. It is statically indeterminate.

STRUCTURES

In Chapter 1 we analyzed the tensions and compressions in the component structural members of the Firth of Forth bridge with the help of a live model. Figure 4.24 shows another truss, of the type used in buildings to span long openings, a so-called *Warren truss*. The bars of the truss are joined together with connections that approximate a hinge. This means that at the joint they can exert only a push or a pull in the direction of the bar, depending on whether the bar is in compression or tension. Let us try to find out which of the bars of the Warren truss in Fig. 4.25 are in tension and which are in compression, without actually stopping to calculate their magnitudes. Of course we now use the method of statics rather than a live model.

First isolate the entire truss as a free body. The body is in equilibrium. Taking moments about joint 1 will give us reaction **Q** (20 kilonewtons) and taking moments about joint 12 at the other end of the truss will disclose the magnitude of the reaction **P** (40 kilonewtons). No information will be gained about the forces in the bars. They remain internal forces, pulling or pushing equally at both end joints. The two equal and opposite forces cancel each

Figure 4.24 Part of a long-span Warren truss supporting the roof structure of a building.

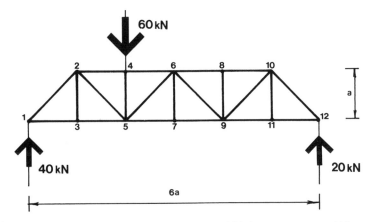

Figure 4.25 Diagram of a Warren truss. All joints are assumed hinged.

other out and have no effect on the equilibrium of the truss as a whole.

To find the internal forces we follow the path indicated by Galileo. We may start in the middle. Make an imaginary cut across line *a–a* and isolate as a free body only the portion to the right of the cut (Fig. 4.26). The body thus isolated is in equilibrium, even though it is being pushed up by a force of 20 kilonewtons at one end. This state of rest is possible only because the body is supported by the three bars 4-6, 5-6 and 5-7. In the free-body diagram, the bars are represented as forces: they are external supports in contact with the free body.

What is the force **D** in bar 5-6? The reaction **Q** pushes up on the body. For equilibrium the vertical component of the force in the diagonal must pull down (neither horizontal bar can do that). Since the bar pulls on the joint to which it is attached, it must be in tension. The forces in the horizontal bars can be obtained just as easily. The body is in rotational equilibrium.

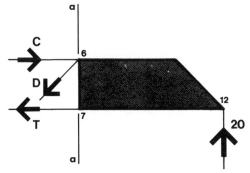

Figure 4.26 The forces in the bars can be found only by applying Galileo's method of imaginary cuts through the structure. Turn the page upside down and note the similarity with the forces in Fig. 4.23.

But the reaction **Q** and the vertical component of the diagonal force form a counterclockwise couple. Their turning effect must be resisted by another couple: force **C** pushing at joint 6 (in compression) and force **T** pulling at joint 7 (in tension).

Note that the moment arm of the horizontal forces is much smaller than that of the vertical forces and so the horizontal forces will have magnitudes considerably larger than that of the reaction **Q.**

A cut through the adjoining panel of the truss will show that the diagonal bar 6-9 must push down to preserve vertical equilibrium (bar in compression), while the forces in the horizontal bars again form a clockwise couple. All the bar forces in the truss can be found by a succession of imaginary cuts through the truss. This is called the *method of sections.*

We may proceed in a different way. The body isolated in the free-body diagram can be any size; it can be half the truss or just a single joint. Figure 4.27 shows the free-body diagram of joint 12. The joint is in equilibrium under the action of the reaction **Q** and the forces in the bars that support it. The magnitudes of the bar forces are readily extracted from the vector triangle where they follow the known reaction and each other tip-to-tail. The diagonal force pushes toward the joint so the bar is a strut. The horizontal force is directed away from the joint—it exerts a pull and the bar must be a tie, working in tension.

We can now move on and isolate joint 11 as a free body. Bar 11–12 exerts the same pull on joint 11 that we found at its other end exerted on joint 12. It is balanced by the forces in the other two bars but, since the horizontal

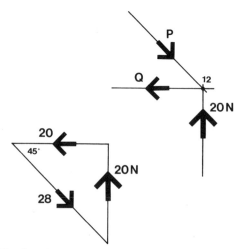

Figure 4.27 The free body can be any size, an entire truss, or just a single joint. The free-body diagram of joint 12 gives the magnitudes of the bar forces at that joint.

pull is completely absorbed by an equal and opposite pull in bar 11–9, bar 11–10 has nothing to do. It is a so-called *zero-force bar.* Similarly, all the vertical bars are zero-force bars, except bar 4–5, which pushes up with a force equal to the vertical load. In theory these zero-force bars may be left out, but in practice they have the job of holding up the long horizontal bars that would otherwise bend under their own weight.

Next we can isolate joint 10 as a free body. The push in bar 10-12 is known and, by means of a vector triangle, we find that the diagonal pulls down while the horizontal bar pushes against the joint. And so on, from joint to joint.

This so-called *method of joints* is a comparatively recent development but, it would seem, familiar to Simon Stevin. Isolate, for example, knot C of the rope shown in Fig. 4.28 as a free body. The knot is in equilibrium under the action of three forces: the pulls of the two sections of the rope and the load **H,** the magnitudes of the forces in the rope are readily obtained by vector triangle, as seen. In Stevin's diagram, triangle *BPC* (similar to our vector triangle) does the same job for knot C, while triangle *MNB* gives

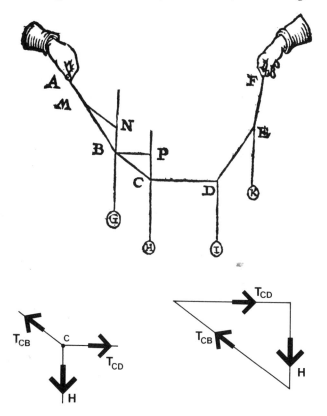

Figure 4.28   Stevin used the same idea to find the forces in the rope.

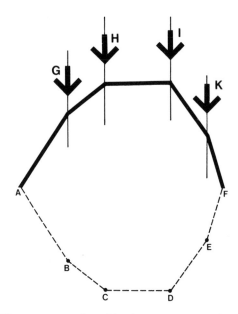

Figure 4.29   The rope turned upside down represents the line of thrust of the loads on it.

the forces in the sections *AB* and *BC* of the rope. The forces in the rest of the rope can be found by additional vector triangles.

The rope supports the loads in pure tension. Imagine the rope made of a rigid material and flipped upside down while the loads remain acting at the same points (Fig. 4.29). Repeating the vector triangles we would find that the structure now supports the loads in pure compression. Thus the inverted rope represents the ideal shape of an *arch* supporting the same loads—it is called the *line of thrust* of the loads. An arch of any other shape would be subjected to bending in addition to compression. It follows that the classical semicircular arch is not a structural shape. When the arch supports a continuous wall above it, the line of thrust has a parabolic shape. The pointed gothic arch comes closer to this ideal shape.

The farther the shape of an arch deviates from the line of thrust, the more bending it has to endure. A typical modern building frame (Fig. 4.30) is essentially a rectangular arch. Maximum bending moments can be expected in the corners and in the middle of the span. A frame made of steel or reinforced concrete is designed to resist this bending. Conformity to the line of thrust is much more important in masonry structures which, because of the nature of the material, are ill suited to resist the tension that accompanies bending. The famous Spanish architect *Antonio Gaudi* designed some of his most daring masonry buildings by constructing a model with strings first and hanging the scaled-down loads from it. He would then fix the

Example   95

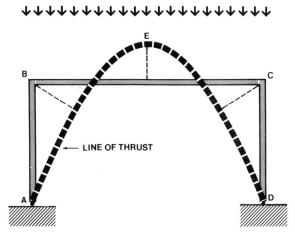

Figure 4.30 A rectangular building frame is an arch that deviates from the parabolic line of thrust of the uniformly distributed loading. Maximum bending will occur at points *B*, *C* and *E* of the frame.

strings and invert the model to find the ideal structural shape for the building.

EXAMPLE

A skyscraper or any tall building acts as a vertical cantilever beam fixed in the ground. The horizontal wind forces bend the cantilever and, to resist the stresses, the structure is designed as a giant tube. The side of the building exposed to the force of the wind and the one on the leeward side become the flanges while the two sides at right angles to the flanges act as a double web of the beam.

The tube corresponds to the typical section of a steel beam. The difference is that instead of a single central web we have two webs on the sides. As the direction of the wind changes, the sides of the building function alternately as flanges or webs. The advantage of this system for tall buildings is that the internal moment arm of the reactive couple is as long as it can possibly be: it is as long as the building is deep. Consequently, the tension and compression forces in the structural members are minimized.

The twin towers of the World Trade Center (Fig. 4.31) in New York are an example of a tubular vertical cantilever. The problem with this solution is that the highly efficient tube must be perforated with openings for windows, which considerably weaken the structure. They have to be kept small, sur-

Figure 4.31 The twin towers of the World Trade Center in New York are giant cantilever tubes.

Ezra Stoller © ESTO

Figure 4.32 The John Hancock Building in Chicago achieves rigidity by means of diagonal bracing, like a truss.

rounded by closely spaced columns and stiff spandrel beams rigidly connected to ensure that each wall acts as much as possible as a stiff plate. The John Hancock building in Chicago (Fig. 4.32) takes a different approach. This building is a vertical truss. The horizontal loads are resolved into tensions and compressions in the diagonals.

CHAPTER FIVE _____

Water and Air

More paradoxes and a horror that is not

_____

THE FORCE OF BUOYANCY

How can a heavy ship (Fig. 5.1) complete with equipment and cargo stay afloat? The discovery of the principle of *buoyancy* is credited to Archimedes who, according to legend, was so excited when he found the answer while immersed in water that he streaked naked from his bath shouting "Eureka! I have found it!"

Two thousand years later, Stevin explained the principle of buoyancy in the following way. Imagine a mass of water D inside a larger mass of the same fluid (Fig. 5.2). Further imagine the surface of the water D solidified so that it surrounds the water inside on all sides. This imaginary container is without mass and infinitely thin. Mass D must be supported by the rest of the water so that it is in equilibrium, or else it would fall or rise. If that was the case it would be continuously replaced by other water filling the gap and we would have perpetual motion, which is just as absurd here as in the problem of the inclined plane. Therefore mass D is at rest inside the total mass of water. Gravity pulls it down with a certain force (the weight of the water D); the surrounding water holds it up with an equal force.

Next pour the water out of the container, so that volume D remains empty. The surrounding water is unaware that anything has happened and it continues to push the container up with the same force as before. If we place inside the container a solid body A of equal volume, obviously the downward pull of gravity on the body will be opposed by the upward push of the surrounding water, equal to the weight of the water that has been taken out.

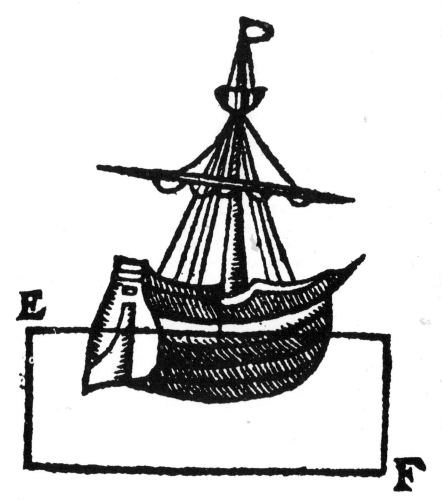

Figure 5.1   The problem of buoyancy: how can a heavy ship stay afloat?

Thus Stevin arrives at the principle of buoyancy, usually called the *principle of Archimedes* after the man who found it first: a body immersed in water is buoyed up with a force equal to the weight of the volume of water displaced by the body. If the density of the body (that is, its mass per unit volume) is equal to that of water, it will float anywhere inside the water; if greater, it will sink. If the body is less dense than water, it will rise to the surface or a force will be required to hold it down.

For example, in Fig. 5.3 Stevin shows a body weighing 12 newtons (in modern units) immersed in water. The density of the body is four times that of water. What is the force **D** required to prevent the body from sinking? The same volume of water would weigh only one-quarter as much as the body,

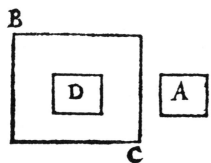

Figure 5.2 A mass of water *D* inside a larger mass of water is in equilibrium. A body *A* of volume equal to *D* would be subjected to the same upthrust as *D*. This force is called buoyancy.

or 3 newtons. The weight of the immersed body will therefore be reduced by 3 newtons and a pull of 9 newtons will suffice to hold it up. If the weight of the body is 2 newtons and its density only one-fifth that of water (Fig. 5.4), a force will be required to keep it down. The same volume of water will

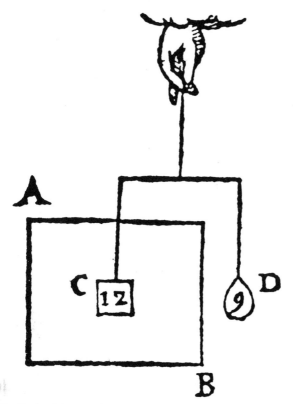

Figure 5.3 Body *C* has a density four times that of water. Only 9 newtons is required to hold it up.

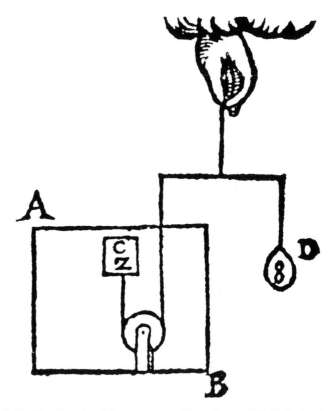

Figure 5.4 The density of the submerged body is one-fifth that of water. A force of 8 newtons is required to hold it down against buoyancy.

weigh $5 \times 2 = 10$ newtons. This will be the force of buoyancy exerted on the body and a counterweight of 8 newtons must be placed on the balance.

A body of a density less than that of water will rise and float on the surface only partially submerged. The portion that remains under will displace a volume of water whose weight will be equal to the weight of the floating body. That is how ships float: even though they may be made of heavy metal, they are hollow and thus their overall density is less than the density of water. When the ship is fully loaded it sinks into the sea deeper than when it is empty.

The modern approach to the question of buoyancy requires far less imagination than Stevin's and creates no impulse to run around shouting *Eureka*. It is just another application of the free-body diagram used to analyze structures and machines. We shall continue to use water in our discussion but, of course, the principle applies to any fluid.

Let us take again Stevin's container full of water (Fig. 5.5). Isolate a volume D of water as a free body. It is in contact with the surrounding water and at every point of contact there is a force pushing on the free body. As yet we

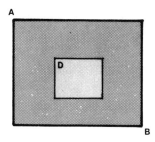

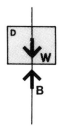

Figure 5.5   The free-body diagram of the mass of water *D*. The resultant of the water pressure—buoyancy—is equal to the weight of the free body.

know nothing about these forces (called *pressure*) except that they must exist. There is another force acting on the body, the pull of gravity **W.** The free body is in equilibrium under the action of the forces, therefore the resultant of the water pressure must be a force equal and opposite to force **W;** furthermore, its line of action must pass through the center of gravity of the volume of water. This resultant is called the *force of buoyancy.*

The rest follows as in Stevin's argument—note how close he came to actually using the idea of a free body. Only one comment can be added. Even if the weight of the immersed body is equal to the weight of the water it displaces, the body will be in equilibrium inside the water only if the center of gravity of the body coincides with that of the displaced water; otherwise the two forces (the weight of the body and the force of buoyancy) will form a couple and turn the body.

THE PRESSURE OF WATER

In his discussion of the principle of buoyancy Stevin resumed the work of Archimedes after an interruption of many centuries. But his work on the pressure of water was a pioneering effort and laid the foundations of *hydro-*

statics, the branch of statics that deals with the equilibrium of fluids. However, before we follow Stevin we must deal with some basic points he did not clearly state.

What exactly do we mean by pressure? The simplest and most accurate definition is a mathematical one: pressure is magnitude of force divided by the area on which it acts:

$$p = P/A$$

Suppose we have a box weighing 10 newtons sitting on a table (Fig. 5.6). The area of contact is, let us say, 0.2 m². The pressure will be equal to

$$p = 10/0.2 = 50 \text{ N/m}^2$$

over the entire area of contact. Just like force, pressure always acts between bodies in contact and is exerted on both bodies. The box presses on the table; the table returns an equal and opposite pressure on the box.

Figure 5.6 Pressure is force divided by area.

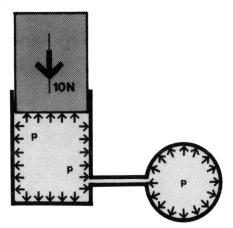

Figure 5.7   Pressure in a closed container is transmitted undiminished through-out the fluid and acts at right angles to all surfaces.

Let us now take our box from the table and place it in a container filled with water. The container fits tightly around our box but there is no friction between the two. The water in the container is now under pressure. The area of contact is the same as before, so the pressure is also the same.

Water, like all other fluids, has the special property that it cannot retain its shape. This is the only property that distinguishes a fluid from a solid. There are two important consequences. First, because of this property the pressure applied to water is distributed undiminished in all directions within the entire volume of water. As shown in Fig. 5.7, it acts on the sides as well as the top and bottom of the container. If a second container is attached to the first one by a pipe, the water in it (and in the pipe) will be under the same pressure as the water in the first container. Second, the pressure of water is always directed at right angles to the surface of the container enclosing it. If we drill holes in the sides of the round container, water will squirt out in all directions with equal force.

The two consequences just mentioned are known as the *principle of Pascal*, after Blaise Pascal, who, in the century after Stevin, gave hydrostatics much of its modern form. Let us state Pascal's principle again, in one sentence: pressure exerted at any place on a fluid in a closed container is transmitted undiminished throughout the fluid and acts at right angles to all surfaces.

Thus far we have neglected to consider the effect the pull of gravity on the water (that is, its weight) has on the pressure inside the container. We now remove the box pressing on the water in our container and look at the effect of weight separately (Fig. 5.8). Suppose that the weight of the water in

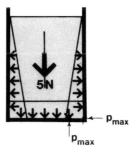

Figure 5.8 The pressure due to the weight of the water itself increases with depth. It is a maximum at the bottom.

the container is 5 newtons. The effect on the bottom of the container is the same as if a solid box of that weight was sitting on it. The pressure is

$$p = 5/0.2 = 25 \text{ N/m}^2$$

The water at bottom level is under the same pressure as the container bottom itself. Since it transmits the pressure in all directions, the sides of the container at that level are under the same pressure too.

At half depth, the pressure is only half the pressure at the bottom; at the top surface it is zero. The pressure on the container changes in the same way, from zero to a maximum at the bottom. Incidentally, this gives us another explanation of the phenomenon of buoyancy. The pressure on a submerged object is at right angles to its surfaces. It pushes down on the top surface and up on the bottom surface but since the bottom surface is deeper, the pressure on it pushing upward is greater than the downward pressure on the top surface. The pressures on the sides of the body are equal and opposite and cancel out.

It is an interesting and surprising fact that the pressure at any level in the water does not depend at all on the amount of water in the container but only on the *head of water* above. Suppose we have two containers filled with water to the same depth h. The area of the base of container 2 is twice that of container 1. The weight of water in the second container, being proportional to its volume ($V = A \times h$), is double the weight contained in the first. However, it is spread over twice the area, therefore the pressure ($p = W/A$) is the same. The pressure on the sides of the two containers will also be the same.

On a larger scale, the pressure exerted on a dam (Fig. 5.9), so long as the depth of water retained by it does not change, remains the same regardless of whether the body of water behind it is a narrow ditch or an ocean.

Figure 5.9 A dam in Quebec, Canada: the pressure depends only on the depth of the water behind the dam, not on the amount.

THE HYDROSTATIC PARADOX

Logical reasoning sometimes leads to unexpected conclusions, especially when combined with a fertile imagination. Let us follow Simon Stevin in one of his most brilliant deductions.

Take a container *ABCD* filled with water (Fig. 5.10). The bottom of the

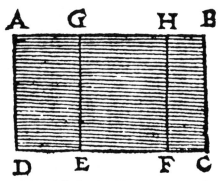

Figure 5.10 A container full of water. The pressure on the surface *EF* is caused by the weight of the water *GHFE* above it.

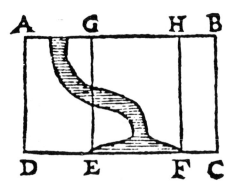

Figure 5.11 Imagine the water solidifed except for what is left in the shaded tube. The pressure on *EF* is unchanged.

container is under uniform pressure; in particular, the pressure on the area *EF* of the bottom is caused by a volume of water directly above *EF* (volume *GHFE*). So far so good. Next, Stevin imagines all the water in the container *ABCD* solidified so that only the shaded tube of water is left ending in the area *EF* on the bottom of the container (Fig. 5.11). The solid part is of the same density as water so that its presence makes no difference to the pressure on *EF*, which remains the same as when there is only water in the container. But if we pour out the water there will be no pressure on *EF*. The startling conclusion is that the water in the narrow tube causes the same pressure on *EF* that the much larger volume *GHFE* directly above it caused before.

In Fig. 5.12 the solid portion takes a different shape so that the tube also changes. Still, logically this makes no difference and the pressure on *EF* must be the same as if the container had only water in it. It is less than before because the surface is less deep but it is not influenced at all by the shape of the tube of water actually exerting the pressure. Figure 5.12 proves that the same pressure is also exerted upward: if it was different, the solid *GHFE*, weighing the same as an equal volume of water, would be buoyed up or it would sink.

Is it difficult to believe that a narrow tube of water can exert the same pressure as a large volume of water. Did Stevin take logic too far? He proves the correctness of his conclusions with a very simple experiment. A narrow tube and a large vessel (Fig. 5.13) are connected across the area *DC*. If we pour in water its surface will be at the same level in both vessels. There can be no doubt about that. However, this everyday experience proves that the pressure is the same from both sides at the common surface *CD*; otherwise the water on the side of the weaker pressure would be pushed up. The pressure therefore does not depend on the amount of water in the vessel. It does depend on the depth, though: pour more water into the thin tube and

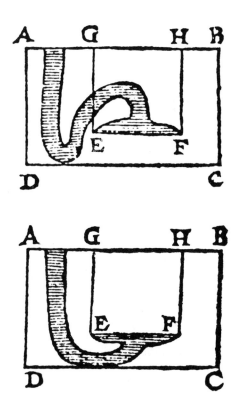

Figure 5.12 The pressure on *EF* is less but still the same as when the container is full of water. The same pressure is exerted upward.

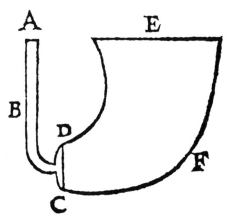

Figure 5.13 The hydrostatic paradox: the small amount of water in the narrow tube balances the large amount in the wide vessel.

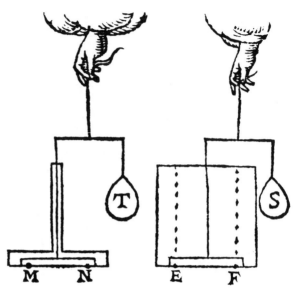

Figure 5.14 The head of water is the same in both containers, therefore the same force is required to pull up the identical disks *MN* and *EF*: *T* = *S*.

its increased pressure will actually push up the larger volume of water on the other side. Equilibrium will be restored when the water is once again at the same level on both sides. This is known as the *hydrostatic paradox*.

Stevin suggests a number of ingenious experiments to illustrate the paradox. For example, take two identical disks of negligible weight covering openings *MN* and *EF* in two vessels of very different volumes but filled with water to the same depth, as in Fig. 5.14. The pressure on the two disks must be the same, and, since they are identical, the force **T** required to pull the first disk up against the small amount of water in the vessel on the left will be equal to force **S** required to lift the larger volume in the vessel on the right.

Let us make another experiment. Take our container of Fig. 5.8 where the pressure of water on the bottom was found to be 25 N/m². In Fig. 5.15 the

Figure 5.15 The upward push *P* must balance the weight of the water in the container, therefore *P* = 5 newtons.

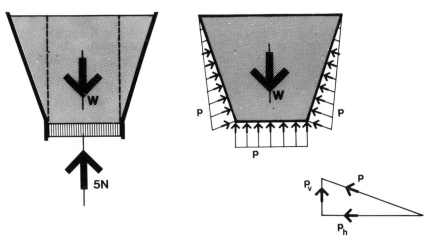

Figure 5.16   The head of water is the same in the large container. It follows that the same force of 5 newtons will hold the piston in place. The free-body diagram contains a clue to the hydrostatic paradox.

bottom of the vessel has been replaced by a movable piston. The magnitude of force **P** required to hold it up is obtained by reorganizing the formula for pressure $(p = P/A)$ to get $P = p \times A$ and

$$P = 25 \times 0.2 = 5 \text{ newtons}$$

It is equal to the weight of the water in the container—which makes sense. Suppose now that we have a much wider vessel filled with water to the same depth but with a piston of the same area as in the previous vessel closing the opening at the bottom (Fig. 5.16). Then the pressure of water on the piston is still 25 N/m² and the same push $P = 5$ N is required to hold up the larger amount of water.

The free-body diagram can be used to shed some extra light on this paradox, which, even though it follows logically from the basic principles of hydrostatics, seems to contradict common sense. Draw a free-body diagram of the water in the wide vessel. The water presses against the vessel at every point of contact with a force directed at right angles to the surface of the vessel. The vessel responds with equal and opposite forces. This pressure exerted on the water by the vessel we show in the free-body diagram in addition to the weight force **W.**

We find that the pressure exerted by the sides of the vessel on the water is slanted upward. The vertical components of the pressure (p_v) assist the piston in holding up the water. When the walls of the vessel are vertical the pressure is purely horizontal and makes no difference to the balance of

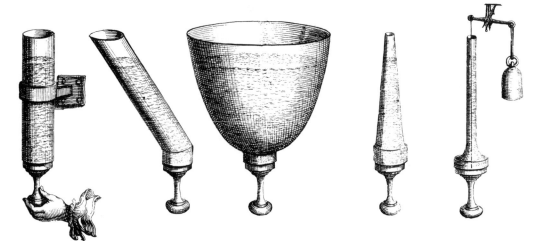

Figure 5.17 Pascal's vases. Could the same force hold up the water in all these very different vessels?

vertical forces. In Stevin's example (Fig. 5.14) the top surface of the container on the left pushes down on the water so that the water between the disk and the container acts like a compressed spring. Pascal made a whole series of vessels, all of different shapes but closed with identical pistons at the bottom (Fig. 5.17). When filled with water to the same depth, they all require the same force to hold up the piston. The force is equal to the weight of the water in the first vessel.

THE HYDRAULIC MACHINE

The two connected vessels of Fig. 5.13 are familiar to most people in one form or another. We accept easily that the water in both of them will be at the same level. Stevin used this experience to confirm his ideas on the pressure of water. A century later, Blaise Pascal went a step further: he saw in the two connected tubes a new hydraulic machine.

Pascal took two vessels, *A* and *B*, and made the cross section of vessel *B* 100 times that of vessel *A* (Fig. 5.18). The vessels are filled with water up to level *N-N*. Suppose the weight of the water in the narrow vessel between levels *O-O* and *N-N* is 10 newtons; then the weight of the water between the same levels in the wide vessel is 100 times greater, that is, 1000 newtons. Imagine this water solidified: we now have a solid of 10 newtons in the narrow tube holding up a solid weighing 100 times as much in the wide vessel.

Pascal realized that the weight forces in the tubes may be replaced by equal forces exerted in a different way without disturbing the equilibrium

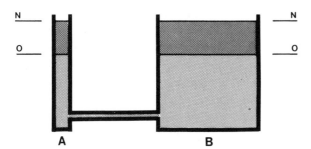

Figure 5.18   Imagine the water between levels *N-N* and *O-O* solidified.

(Fig. 5.19). "If a perfectly fitting piston is adjusted to each tube," wrote Pascal, "then one man pressing on the smaller piston will exert a force equal to that of one hundred men pressing on the larger, and will exceed that of ninety-nine men doing the same." A push of 10 newtons on the small piston will balance a force of 1000 newtons on the larger one, just as two forces on a lever are balanced when its arms are in the ratio of 1 : 100.

A change in the ratio of the areas of the pistons may multiply the force exerted on the smaller piston even more. "Whence it follows," continued Pascal, "that a vessel filled with water is a new mechanical principle and a new machine that will multiply forces to any amount desired; for a man by this means will be enabled to lift any weight that another may propose."

Pascal saw the cause of this multiplication of force in the pressure exerted by the pistons on the water. Suppose the cross section of tube *A* is 0.01 m² and that of tube *B* is 100 times larger, or 1 m². A force of 10 newtons in the first tube will create a pressure in the water

$$p = 10/0.01 = 1000 \text{ N/m}^2$$

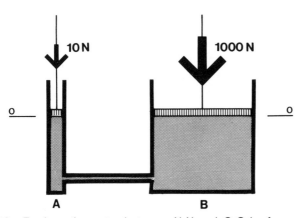

Figure 5.19   Replace the water between *N-N* and *O-O* by forces exerted in some other way. The connected vessels function as a hydraulic machine.

and the force on the larger piston, equal to 1000 newtons but acting on an area of 1 m², will cause the same pressure. Thus the pressure in the water is the same throughout and the machine is in equilibrium. Pascal added a warning: the pressure is transmitted undiminished in all directions and therefore all parts of the machine, including the connecting pipe, must be strong enough to resist it.

The great advantage of the hydraulic machine over machines based on the lever and the inclined plane is its great flexibility. The two connected tubes may be a long distance apart with the pressure transmitted through a simple pipe. Still, Pascal's invention did not see a practical application for more than a century. Its time came on the eve of the industrial revolution (in 1795, to be exact) when it was reinvented and patented by *Joseph Bramah*, a Yorkshire farmer's son who became a versatile inventor. (Among his other inventions were an improved water closet and a lock that nobody managed to pick for 67 years despite a standing challenge prize.)

The hydraulic machine has played an essential part in industry ever since. In the nineteenth century the city of London had a hydraulic system that delivered pressurized water through street mains to drive machines in factories, hoists, cranes, lift bridges and so on. Probably the most common modern application is the hydraulic jack. It is known that the Italian architect Domenico Fontana, during the Renaissance, used 40 capstans worked by 960 men and 75 horses to lift an obelisk in Rome. When a similar obelisk was raised in London four centuries later, four hydraulic jacks were used, each worked by one man.

Hydraulic pistons are used in many applications where large forces are required and where the transmission of forces takes a roundabout route. Modern automobiles and aircraft use them to operate their brakes. The typical tower crane (such as we saw in Chapter 2) has the ability to grow with the building it services: it pushes itself up when required with the help of hydraulic jacks. A single hydraulic piston in Fig. 5.20 raises the amusement park wheel filled with people.

Apart from his invention of the hydraulic machine, Pascal's main advance on Stevin was his explanation of a phenomenon that Stevin never touched: nature's apparent abhorrence of the vacuum that had remained firmly entrenched as an answer to a series of problems ever since Aristotle first established it in the fourth century *B.C.*

THE HORROR OF THE VACUUM

Questions about the nature of the air we breathe and whether it has weight have occupied people since Antiquity. Until the seventeenth century it was

Figure 5.20 A hydraulic ram lifting the spinning wheel of an amusement park machine. Note that the moment arm of the force is shorter than that of the load—a huge force is required!

generally believed that air did not have weight and some even claimed that it had "lightness" rather than weight since it tended to rise.

A seemingly unrelated set of problems was considered solved in a perfectly satisfactory manner by Aristotle and his followers. Why does a pump draw water? Why does water rise and follow the plunger of the pump as if it adhered to it? These and a lot of other questions were explained as due to nature's "horror" of the vacuum that would be left when the piston was withdrawn if water did not take its place. It was noticed, however, that suction pumps refused to work over a height greater than about 10 meters. Galileo believed that this was actually a measure of how much nature abhors a vacuum: he compared the column of water in the pump to a metal rod suspended from its upper end which may be lengthened until it finally breaks under its own weight.

In 1644 Galileo's pupil *Evangelista Torricelli* performed a crucial experiment. He took a glass tube about a meter long and sealed one end. He filled the tube with mercury, placed a finger over the open end so that no mercury could escape and turned the tube upside down with the open end in a bowl of mercury (Fig. 5.21). When he removed his finger the mercury sank at once to about 0.76 meter above the level of the mercury in the bowl, leaving the top of the tube apparently empty—a vacuum. The level of mercury remained exactly the same when the experiment was repeated with several different tubes (Fig. 5.22).

Torricelli was convinced that it was the pressure of air that was holding up the column of mercury in the tube as well as pushing up the water in

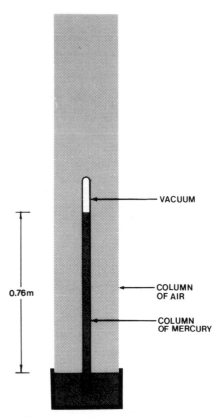

Figure 5.21   Torricelli's experiment: a short column of mercury balancing a very tall column of air.

the pump. The pressure at the surface of the fluid in the bowl is the same at all points on the surface, or the surface would not be level. It is therefore the same under the column of mercury and under the column of air. Since mercury is about 13.6 times denser than water the same air pressure that can sustain a column of mercury 0.76 meter high could obviously push water up to a height of about $13.6 \times 0.76 = 10$ meters. *Robert Boyle* in England tested this conclusion in an impressive full-scale experiment (Fig. 5.23), as had Pascal before him.

Torricelli's experiment caused a great stir and much controversy. The horror of the vacuum was not abandoned by its proponents without a fight. Blaise Pascal dealt it a final blow by a series of experiments that established a complete analogy between the phenomena conditioned by water pressure and those due to air pressure. For example, he showed that mercury, as a result of water pressure, rises into a space containing no water just as it rises into a space containing no air due to air pressure. He attached a bag of

Figure 5.22 The length and shape of the tube above the mercury make no difference.

mercury to a tube open at the top. The deeper the bag was sunk into water, the higher the mercury rose in the tube due to the increasing pressure.

"We live at the bottom of an ocean of air," Torricelli had said. Pascal devised an experiment to show that if we move closer to the surface of this ocean the atmospheric pressure is reduced. In 1648 he sent his brother-in-law to the top of a mountain with a glass tube and a bowl of mercury. The effect was the same as raising the bag of mercury higher in the water: the mercury in the tube sank lower. Many years later, when the first expedition managed to climb Mt. Everest, they had a similar apparatus with them; at

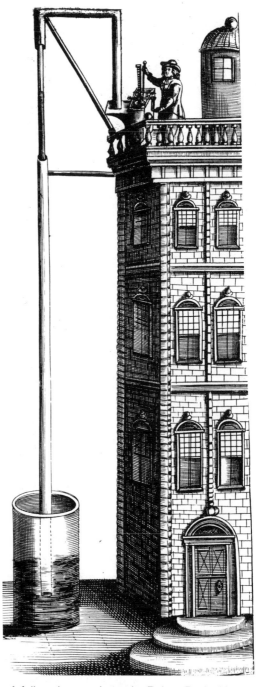

Figure 5.23 A full-scale experiment by Robert Boyle. Vacuum at the top of the tube was created by means of an air pump.

the peak, about 8 kilometers high, the column of mercury measured only 0.27 meter.

Pascal finally had the proof he needed. He wrote: "It is quite certain that there is much more air that presses on the foot of a mountain than there is on its summit, but one cannot well say that nature abhors a vacuum more at the foot of the mountain than at its summit." He was still not completely satisfied. In another, most ingenious experiment he showed what would happen if we could rise to the surface of this ocean of air. He proved that when the pressure of air is completely absent, mercury does not rise in the tube at all.

For this experiment Pascal had a glass tube blown and bent to leave two vertical legs, each about a meter long with a large bulb between the two (Fig. 5.24—but note a mistake in the original illustration: the bulb was really made so that it was completely filled with mercury with no air passage at B). Pascal closed with a finger the hole left at the top of the lower leg.

When the entire apparatus was filled and erect in a bowl of mercury, the lower leg functioned as an ordinary tube did before. There was no air pressure in it and mercury stood at the height of 0.76 meter. No atmosphere

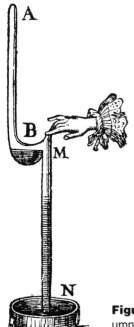

Figure 5.24 When the finger is removed, the lower column of mercury falls in the bowl and mercury rises in the upper tube. There should be no air passage at B.

pressed down on the surface of mercury in the bulb and the mercury did not rise in the upper tube at all. Finally, Pascal slowly opened the hole at the top of the lower leg to admit air. As the air entered, the mercury in the lower leg dropped into the bowl while that in the upper leg rose until it reached a height of 0.76 meter.

THE ATMOSPHERIC ENGINE

Pascal's work on air pressure was paralleled in Germany around 1650 by the mayor of Magdeburg, *Otto von Guericke*, who, working independently, arrived at similar insights. Guericke was determined to create a vacuum and, realizing that air was a fluid like water, he redesigned a conventional water pump so that it could pump air out of a closed vessel. With this pump he performed a series of experiments. In one of them, which became famous throughout Europe, teams of horses could not separate two close fitting brass hemispheres placed together to form a sphere (Fig. 5.25), after Guericke had pumped out the air.

Less spectacular but probably more important were Guericke's later experiments. He attached a metal cylinder to a firm support, with its open end upward (Fig. 5.26). A close-fitting piston was held against stops at the top of the cylinder by a rope passing over a pulley. A group of people firmly held the end of the rope. Guericke then pumped air out of the cylinder through a valve at the bottom and, when a partial vacuum had been produced, the pressure of the atmosphere pushed the piston down into the cylinder in spite of the best efforts of the people holding the rope. The force produced in this way amazed the spectators. Knowing the area of the piston used in the experiment (0.13 m²) we can easily calculate what it was. The pressure of the air is about 100 kN/m². The force exerted by the air on the piston must then be

$$P = p \times A = 100 \times 0.13 = 13 \text{ kilonewtons.}$$

Of course, this force could be achieved by the piston only under ideal conditions. The actual force must have been less.

This experiment suggests ways in which the pressure of the air could be used as a new source of power, sorely needed by the industry of the time. The inventor of the first such engine was one of the leading members of the newly founded Academy of Sciences in Paris, *Christian Huygens*. In 1673 Huygens constructed an apparatus similar to that used by Guericke but, instead of patiently pumping out the air, he decided to take a short cut by using a small charge of gunpowder in the cylinder (Fig. 5.27). The explosion

Figure 5.25 Two teams of horses struggling unsuccessfuly against air pressure at Magdeburg.

Figure 5.26   A group of Magdeburgers opposing air pressure with no success.

of the powder in the capsule *C* would force most of the air out of the cylinder through a nonreturn valve *EF* (simply a hose made of moist leather). After the gases left inside had cooled down, their pressure would be very low and the piston *D* would be pushed down by the atmospheric pressure.

The experiment was successful. A very small charge of gunpowder was exploded in a cylinder 0.3 meter in diameter. It lifted a load of about 700 newtons through a height of 1.5 meters. Thus was born the first *atmospheric engine* and, some people believed, the era of peaceable use of gunpowder was ushered in. However, great expectations on both counts proved too optimistic. "The discovery can serve all sorts of needs where it is required to combine great power with lightness, as in flying, which can no longer be rejected as impossible, although its realization will still require much more work," wrote Huygens of his engine. But gunpowder proved impractical and dangerous as a source of power. A different type of internal combustion engine had to be invented before powered flight could be attempted.

In the meantime, it occurred to Huygens' assistant, *Denis Papin*, that a more efficient method of creating a vacuum inside the cylinder might be by the condensation of steam. In 1690 Papin put a small quantity of water in the cylinder and heated it until the water boiled. The steam pressure forced

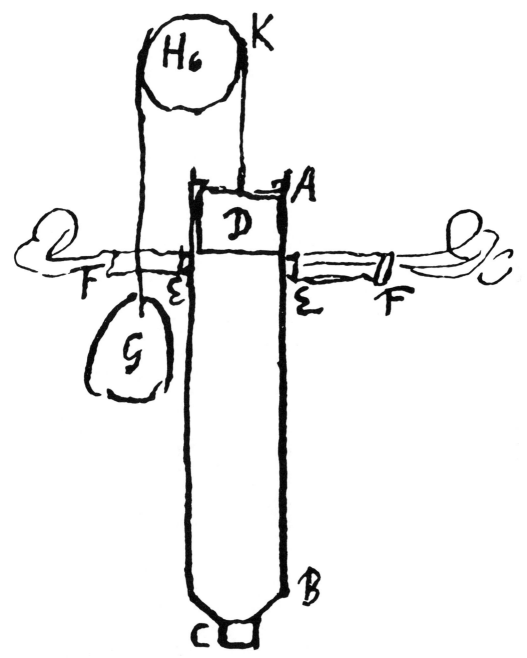

Figure 5.27 Gunpowder for peace: the atmospheric engine of Christian Huygens.

the piston up until it reached the stops at the top. The flame was then removed and the cylinder was allowed to cool. As the steam condensed the pressure in the cylinder became very low and the piston was pushed down by the pressure of the atmosphere as before. Using a small cylinder Papin was able to make a load go up and down by successively applying and removing the source of heat.

Like Huygens, Papin envisaged great possibilities for his invention. It could be applied to extract water from the mines, to sail against the wind—and to throw bombs. This engine did not prove very practical either and Papin remains best known as the inventor of a kind of pressure-cooker.

Still, the basic idea of using steam was sound and eventually it led to the invention of the steam engine. But neither Papin's engine nor its early developments by Newcomen and others could be called steam engines. In all of them, steam was used only as a means of creating a vacuum while the real work of lifting the load was performed by the pressure of the atmosphere. In the steam engine proper, developed by James Watt, the pressure of steam is used to drive the piston both ways. In this form it revolutionized industry and transport (Fig. 5.28).

Figure 5.28 A century later steam replaces gunpowder and reigns supreme for another hundred and fifty years.

Example   125

EXAMPLE

A lift lock replaces a series of conventional locks where boats have to be raised and lowered considerable heights within a relatively short distance. The lift lock at Kirkfield, Canada, raises and lowers boats in two water-filled chambers through a vertical distance of 15 meters (Fig. 5.29). Each chamber

Figure 5.29 The lift lock: two water-filled chambers go up and down like the pans of a balance—with or without boats inside.

Figure 5.29   *(Continued)*

is supported by a huge hydraulic ram, 2.28 meters in diameter, similar to those used to raise cars in service stations. The two chambers are arranged so that one is at the upper level when the other is at the lower. When they are in place, gates are lowered to connect them with the canal; when the boats are safely inside, the gates are raised, sealing off the water-filled chambers.

The upper chamber stops approximately 300 millimeters below the water level of the upper canal and takes on about 130 cubic meters of extra water. The two rams are connected by a pipe and operate as a closed hydraulic system so that the weight of the extra water in the upper chamber provides the force required to push up the opposite ram and raise the lower chamber.

The installation was completed in 1907. At that time, a journalist watching the operation of the lift lock described it as "a wonderful and massive mechanism by which the simple natural law of equilibrium makes a steamer laden with a hundred passengers rise through the air like a bird, or a barge laden with wheat descend as gently and smoothly as a tuft of down."

Thus the two chambers move up and down like the pans of a balance with only a comparatively small force required to set them in motion. The interesting thing about the lift lock is that the weight of the chambers is the same regardless of whether they are just full of water or containing a "steamer laden with a hundred passengers." The steamer (or any other boat) displaces a quantity of water exactly equal in weight to that of the steamer itself.

CHAPTER SIX

A Different Approach

And some new insights

A COMMON RULE

The method of operation of the simple machine known as the pulley was explained in Chapter 2 by analogy with the lever. Stevin used a different approach, presented very briefly and in modernized form here.

Consider the pulley arrangement in Fig. 6.1. What is the force **F** required to balance and, with just a bit of extra effort, raise the load **B?** Stevin's analysis is based on the fact that the force throughout the length of the rope that goes around the pulleys remains the same, always assuming that the rope is perfectly flexible and there is no friction.

Isolate the movable pulley as a free body. There is a force **B** pulling down on the block, replacing the load. There are three identical rope forces pulling up at the points where we cut the ropes. Since the body is in equilibrium the three rope forces must balance the load and, all three forces being the same, their magnitude will be equal to $F = B/3$. This is the pull that must be exerted by the hand to hold up the load.

Thus the application of the free-body diagram to pulleys means that to find the effort required to balance the load, all we have to do is count the number of rope segments holding up the movable pulley (or pulleys) then divide the load into this number. The magnitude of force **F** in Fig. 6.2 is therefore equal to $B/4$.

The fixed pulley at the top has no other function but to redirect the rope forces since it acts as an equal-arm lever. When an extra fixed pulley is added to the system, as in Fig. 6.3 (marked N), it makes no difference to the

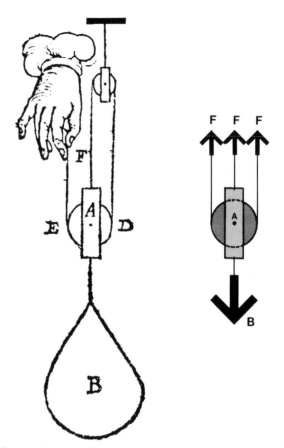

Figure 6.1   A pulley and its free-body diagram: $F = B/3$.

equilibrium of the machine. The movable block of pulleys is still held up by four rope forces and the same force **F** is required at the free end of the rope. The purpose of the extra pulley is purely practical: it is more convenient to pull down on the rope than up. Figure 6.4 shows two pulley blocks of a mobile crane, one with two and the other with four falls of rope supporting the hook. Most tower cranes also allow a rapid conversion from two to four ropes, thereby doubling their lifting capacity.

By adding more pulleys to the system or by combining pulleys in different ways we can produce any degree of multiplication of force. In theory at least we can lift any load by means of a force that can be as small as we like. But are we getting something for nothing? The answer to this question is contained in what Stevin calls a "common rule," which he mentions in passing while talking about pulleys. It says, in essence: *what is gained in force is lost in distance moved.*

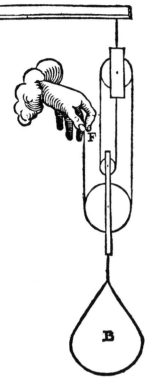

Figure 6.2 Just count the ropes holding up the movable block of pulleys: in this case $F = B/4$.

The truth of this rule is easy to verify. Take, for example, the pulleys in Fig. 6.3. If the load **B** is to be lifted through a distance d, each of the ropes holding up the movable block must be shortened by as much as the load rises, that is, by a length d. This shortening must come out at the end of the rope where the effort is applied by the hand. The end of the rope must be pulled through a distance $4 \times d$ or four times farther than the load is lifted. But we already know that, since the load is carried by four rope segments, the effort required to lift it is equal to $B/4$. We put into the machine a small force through a large distance and take out a large force through a proportionately smaller distance. What is gained in force is lost in distance moved, as the rule states.

As another example let us take a slightly more complex combination of pulleys (Fig. 6.5). Suppose the load **B** of 900 newtons is lifted a distance of 10 millimeters. To accomplish this, each of the ropes from which the lowest movable pulley C hangs must be shortened by 10 millimeters and the upper movable pulley D, which pulls up on the end of the lowest rope, must be raised 30 millimeters. But to raise this pulley 30 millimeters each of its three

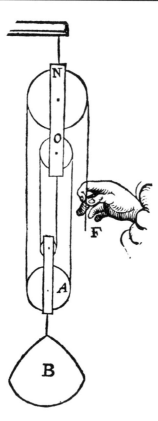

Figure 6.3   Ignore the fixed pulleys. They only change the direction of the pull: $F = B/4$.

supporting ropes must be made shorter by that amount and the force **F** at the end of the rope will have to move a distance of 90 millimeters.

What is the magnitude of force **F**? Draw the free-body diagrams of the two movable pulleys (Fig. 6.6). What we have here is the pulley of Fig. 6.1 repeated twice. The force in the rope on which the lower pulley hangs is equal to $B/3$, or 300 newtons. This same force is shown pulling down on the upper movable pulley in the second free-body diagram (the two pulleys, via the rope, pull on each other with equal and opposite force). Then the force **F** in the rope holding up the upper pulley is equal to only 100 newtons, or $B/9$. We find that the load of 900 newtons moves through a distance of 10 millimeters while the effort, equal to only 100 newtons, traverses a distance of 90 millimeters.

Stevin did not pay much attention to his common rule. It seems he looked upon it as a practical means of assessing simple machines, familiar to most people but unworthy of philosophers. There is much more to it, however. Duly refined, it now serves as basis for a statics which does not need the concepts of moment and triangle of forces.

Figure 6.4 A pulley block supported by four falls of rope instead of two can be used in cranes for extra lifting capacity.

THE WORK OF FORCES

The common rule that Stevin mentioned in connection with pulleys can be extended to other simple machines. Consider the familiar lever (Fig. 6.7). If its arms are in the ratio of 4 to 1, then the load **B** will be four times larger

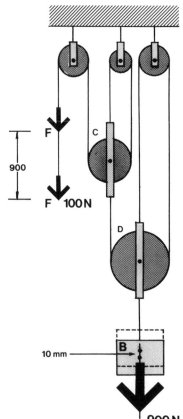

Figure 6.5 The load moves up only 10 millimeters while force **F** pulls through a distance of
900 millimeters.

than the effort **F**. But it can be easily shown by geometry that if the load is
raised a distance *d*, then the distance *h* through which the other end of the
lever must move is equal to 4*d*.

Stevin did not take this any further, but Galileo did. He observed that the
product of force and distance moved is the same at both ends of the lever:

$$B \times d = F \times h$$

The same applies to all the pulleys in the previous examples and, as we
shall see, to all other machines. In view of its importance, this product of
force and distance is now given a special name: *work*. A force pulls or
pushes a body in a certain direction. If it actually manages to displace the
body in that direction, it has performed work and the amount of work thus
performed is measured by the product

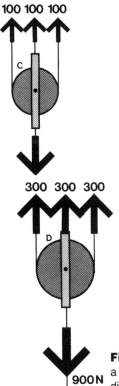

100 100 100

300 300 300

900 N

Figure 6.6 Only 100 newtons of force is required to lift a load of 900 newtons. What is gained in force is lost in distance moved and vice versa.

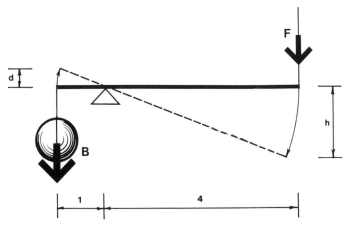

Figure 6.7 The common rule applies to the lever.

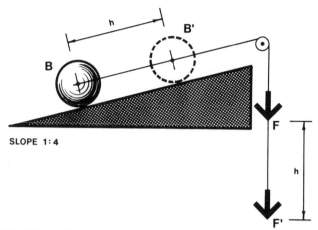

Figure 6.8 The inclined plane is no exception but we must take a closer look at displacements and forces.

$$\text{work} = \text{force} \times \text{displacement}$$

The basic units of work are the same as those of moment: newton-meters.

A look at the inclined plane will clarify the concept of work further (Fig. 6.8). At first glance, it seems that here we have a discrepancy. The effort **F** and the load **B** move the same distance h even though, since the slope of the plane is 1 : 4, the effort required to balance the load is only $B/4$ according to the law of the inclined plane. Galileo understood perfectly well what was happening in this problem: the load itself indeed moved along the plane a distance h, equal to that traveled by the effort. But in the process it was raised *vertically* only a distance $d = h/4$ and again the product of force and displacement in the direction of the force is the same for both forces.

The force acting on the load is the force of gravity and only the vertical displacement in the direction of the force counts. This is true in all cases. If the body on which a force acts moves but not in the direction of the force,

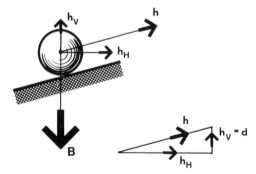

Figure 6.9 Displacements are vectors. Only the component in the direction of the force counts.

the force cannot get full credit for the displacement. Only the displacement in the direction of the push is taken into account (Fig. 6.9). Displacement is a vector—it has magnitude and direction—and it can be resolved into components using vector algebra. Only the component in the direction of the force counts. Furthermore, even though work is the product of two vectors (force and displacement) it is itself a scalar quantity. It has only magnitude, not direction.

The common rule of Stevin can now be expressed in terms of work. We get out of a machine exactly as much work as we put in. The machine simply changes the ratio of the two factors that constitute work. At one end we put in more displacement and less force, at the other end we take out the same amount of work, but in the form of more force and proportionately less displacement. It takes as much work to lift a load with a machine as without it. The advantage of the machine is that it enables us to perform work that unaided we could not perform; it does not create work. Once again we find that there are no "free gifts" in nature.

The famous French philosopher and mathematician *René Descartes* (Fig. 6.10) saw in this a fundamental principle of nature which explains not only under what circumstances forces balance on a simple machine but also *why* they are in equilibrium. It is evidently the same thing, says Descartes, to lift a load of 100 newtons to the height of 2 meters or to lift 200 newtons to a height of only 1 meter or 50 newtons to 4 meters, and so on. For example, look again at Fig. 6.2. The load **B** is raised through a distance d; the hand pulls the end of the rope through a distance $4 \times d$ but the force required is $B/4$. We could just as well have cut the load into quarters and lifted each quarter through a distance d. The net result—the total work performed—would be the same:

$$B \times d = B/4 \times 4d$$

The machine is needed when we cannot cut the load into manageable smaller pieces.

Common sense dictates also that only the component of the displacement in the direction of the force counts when we calculate the work performed by a force. For example, if our objective is to raise a large rock to a height of 2 meters we may make the task easier by placing the load in a wheelbarrow and pushing it 10 meters along a ramp to one side and then another 10 meters back along another ramp to where we want it. However, a stronger worker may be satisfied with a shorter but steeper ramp or he may even be able to lift the load straight up. The net result in all three cases is the same: the rock is lifted 2 meters.

Figure 6.10   The philosopher René Descartes.

SIMPLE MACHINES REVISITED

In Chapter 2 we showed the practical advantages of the wheel-and-axle machine and explained its operation by means of the law of the lever. Consider now what happens when the machine is in operation (Fig. 6.11). When the large wheel is rotated through one full revolution the length of rope uncoiled from it is equal to its circumference:

$$h = 2 \times R \times \pi$$

and the counterweight **K,** attached to the rope (which really represents the

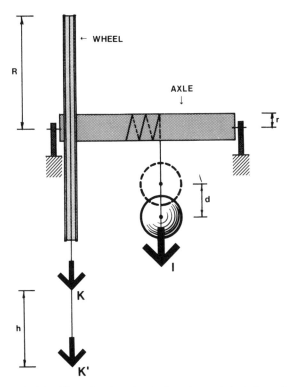

Figure 6.11   The wheel-and-axle from the point of view of work.

effort exerted on the machine), drops that distance. At the same time, the length of rope wound up on the axle is equal to the circumference of the axle:

$$d = 2 \times r \times \pi$$

The load **I** is raised the same distance d. Knowing that the two forces must satisfy the condition of moment equilibrium $K \times R = I \times r$, we can easily calculate that it is also true that

$$K \times h = I \times d$$

or the work done by the effort **K** in pulling down through a distance h is equal to the work done on the load **I** in raising it through the smaller distance d.

The wheel-and-axle winch has a clear advantage over the simple lever, but it also has its own limitations. If we want to increase the effectiveness of this machine we can do it only in one of two ways: either by increasing the radius R of the wheel or by decreasing the radius r of the axle. If we enlarge

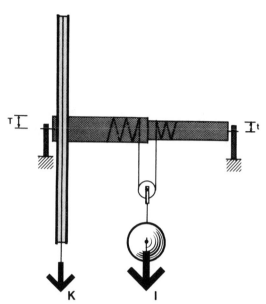

Figure 6.12   An ingenious invention that is far from obvious without the concept of work: the differential axle.

the wheel, the machine becomes unwieldy. If, on the other hand, the thickness of the axle is reduced, its strength may become inadequate for the larger load it is now capable of lifting.

This dilemma was solved in a very ingenious way by using a so-called *differential axle*. The axle is in two parts of different thickness, fixed together (Fig. 6.12). The rope to which the load is attached is coiled on the thicker part of the axle (of radius T) in one direction; it goes down and around a pulley to which the load is attached then up again to the thinner part of the axle (radius t) on which it is coiled in the opposite direction. The result is that as the wheel is turned, the rope winds off one end of the axle and onto the other, seemingly defeating the object of the whole exercise.

The simple but effective operating principle of this machine is based on the work of the forces involved. The difference in the diameters of the two portions of the differential axle is small, but it exists. There is always more rope wound up than released—but very little more. Consequently, the load is raised only a very small amount while the other force, the effort, has to traverse long distances to achieve this small displacement of the load.

By comparing the distances traveled by load and effort, we find out that the differential axle is equivalent to an ordinary axle which has a radius

$$n = T - t$$

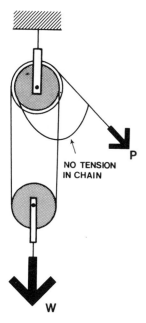

NO TENSION
IN CHAIN

P

W

Figure 6.13   The modern differential hoist is a development of the old differential axle. It is portable and takes up less space.

The power of the machine does not depend on the difference between the diameters of the axle and the wheel, but on the smallness of the difference $(T - t)$. The axle may be made as thick as required for strength and yet the difference $(T - t)$ may be so small as to make the machine capable of any desired multiplication of force.

A modern development of this idea is the *differential pulley*. It is really the same device as the one just described, only the axle has shrunk in length so that it has become two wheels of different diameters that are attached (Fig. 6.13), forming a double pulley. An endless chain is slung twice over the top pulley so that as the force P is applied, the chain is wound up on the bigger wheel (thus raising the load) and at the same time released from the smaller wheel (lowering the load). Of course, the load is always raised a bit more than it is lowered.

The seemingly paradoxical hydraulic machine invented by Pascal can also illustrate the concept of work. The ratio of the areas of the two pistons (A and B) in the machine in Pascal's example is 1 : 100. Then a man pushing down on piston A with a force **P** can balance and raise a 100 times larger load **Q** placed on piston B. But look at the displacements of the two pistons that take place in the process. If the man pushes the piston down a distance d, a volume of water equal to

$$V = (\text{area } A) \times d$$

is pushed into the machine and raises piston B. But since the area of piston B is 100 times larger it moves up only 1/100 of the displacement of piston A to accommodate the extra volume of water pushed into the larger tube. The work of the two forces, **P** and **Q**, is the same.

In connection with this machine Pascal mentions another explanation "which will appeal only to geometricians and may be disregarded by others." The explanation is instructive even to nongeometricians and we examine it in the next section; later we shall see the connection between it and the concept of work.

THE PRINCIPLE OF TORRICELLI

The explanation of the hydraulic machine that Pascal believed would appeal to mathematically minded people is based on a very simple idea: "I assume as a principle," said Pascal, "that a body never moves by its own weight without lowering its center of gravity." In other words, if the center of gravity of a body (where all its mass appears to be concentrated) is supported so that it cannot move down, the body itself will not move down either under the pull of gravity. As was the custom in those days, Pascal does not mention that this principle was first published and used by Torricelli.

Taking this principle as self-evident, Torricelli applied it to machines. He reasoned as follows: when two heavy bodies are connected on a simple machine such as a lever (Fig. 6.14), they are forced to behave as a single body: if one of them moves, the other must move too. They can move independently only if they are disconnected from the machine. If this com-

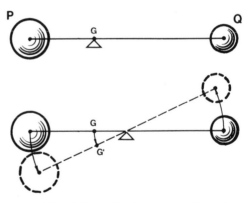

Figure 6.14   The principle of Torricelli: the center of gravity of the composite body cannot move down when the lever is in equilibrium.

posite body is in equilibrium under the pull of gravity, the machine must support it in such a way that its center of gravity cannot fall down. The relative positions of the components may change but the total body, embodied as it were in the center of gravity, does not move downward.

The truth of this statement is easy to see in simple machines. Visualize the lever and the two loads on it as a single rigid body and assume that part P of the body has twice the mass of Q while the lever itself is assumed to have none. The center of gravity of the body will be at point G, which divides the lever in the ratio 1 : 2. This is the point where we have to place the fulcrum. In any other position the fulcrum will not prevent the center of gravity from falling and equilibrium is not possible.

Take the two bodies P and Q and connect them by means of a string into a new composite body supported on an inclined plane (Fig. 6.15). Since $P = 2Q$ the slope of the plane must be

$$BC : AB = 1 : 2$$

according to the law of the inclined plane. The center of gravity G of the composite body divides the connecting line (shown dashed) in the ratio 1 : 2, as before. Now displace the composite body on the plane from its original position. One of its components P goes up, the other one Q goes down. If we draw the connecting line again and divide it in the proper ratio 1 : 2, we find the new position of the center of gravity shifted sideways to G' but not in a vertical direction. This remains so no matter how we move the bodies on the inclined plane, as long as they remain connected and the slope is unchanged. The composite body on the plane is supported in such a way that its center of gravity cannot move down—only sideways.

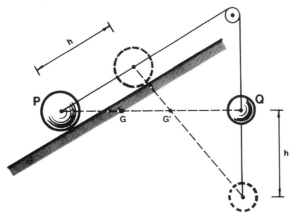

Figure 6.15   The center of gravity of the composite body can only move sideways.

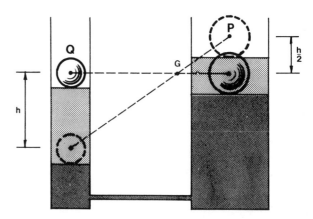

Figure 6.16 The center of gravity of the composite body floating in a hydraulic machine remains in the same spot just as when its two components are attached to a lever.

In Fig. 6.16 the same two bodies P and Q are placed on pistons inside two connected vessels. The cross-sectional areas of the vessels are in the ratio of 1 : 2 and, if we take the weight of the pistons as negligible, the hydraulic machine is in equilibrium. The center of gravity of the composite body divides the connecting line in the inverse proportion of the weights of the two component bodies—in the ratio 2 : 1. When they are displaced from their original position so that Q moves down a distance h and P up half that distance, the center of gravity G of the body will be found at precisely the same point as before the displacement.

This is the explanation of the hydraulic machine that Pascal had in mind. When the two pistons, considered as one system, are displaced up or down their common center of gravity remains stationary. Therefore no movement takes place under the pull of gravity on the pistons and the machine is in equilibrium. It is as though the two pistons were sitting at the ends of a lever whose fulcrum is at point G.

Pascal attached a great deal of importance to this principle. He proudly proclaimed: "I have demonstrated by this method in a little treatise on Mechanics, the reason of all the multiplications of force to be found in all the other mechanical devices invented so far. For I show that in all of them unequal weights, when made to balance with the aid of machines, are so disposed by the machinery that their common center of gravity is not lowered whatever their position; whence it follows that they must remain at rest, that is, in equilibrium." Thus here we have another approach to the statics of machines which, just like the principle of Descartes, aims to replace the law of the lever and that of the inclined plane, or the principle of moment equilibrium and the triangle of forces.

The claims of Descartes and Pascal were perfectly justified. They based their arguments on axioms which are as basic as those that gave us the laws of the lever and inclined plane. At the beginning of the eighteenth century a famous Swiss mathematician, John Bernoulli, gave these principles the form that turned them from explanations of simple machines into generally valid conditions of equilibrium.

THE PRINCIPLE OF VIRTUAL WORK

It was one of Stevin's firm principles never to mix scientific proofs with polemics. He by no means tried to avoid polemics; he just kept them separate in an appendix "in which, among other things, many errors about the qualities of weights are refuted." One of these errors is the attempt to explain equilibrium in terms of displacements that happen when a machine is set in motion—so fundamental to the principle of work. Stevin rejected such reasoning with crushing logic: a body at rest does not displace and where there is no displacement, displacements cannot be used to prove anything.

Sometimes logic can be an obstruction rather than help. With deeper insight, *John Bernoulli* in 1717 made the consideration of forces and their displacements the basis of a new and different approach to statics. There are no displacements when a body is in equilibrium. Stevin was undeniably right in this, but Bernoulli asked himself what would happen *if* a displacement were to happen, a displacement that is only imagined and not caused by the forces acting on the body. Such a displacement is called a *virtual displacement*.

Let us take a simple example. The man in Fig. 6.17 is pushing down his end of the lever with a force **F**; gravity pulls down on the load with a force **B**. Suppose the two forces are balanced; next, imagine the lever rotated so that force **F** goes down a distance h while the load is lifted a corresponding distance d. We found that the work done at one end of the lever is equal to the work done at the other:

$$F \times h = B \times d$$

or

$$F \times h - B \times d = 0$$

The second expression is just another way of writing the first one, but it can be interpreted in a new way.

Force **F** (or any other force) pushes the point on which it acts in its own

Figure 6.17 The total work performed by both man and load is zero in a displacement of the lever.

direction. If it succeeds and that point is displaced from its original position in the direction of the force, the force has performed *positive* work. The load on the lever, force **B,** pulls on the lever, but when the lever is rotated the displacement of the point on which force **B** acts is in a direction opposite to the pull of the force. The force still performs work but this time opposing the displacement. That counts as *negative* work. Thus when the two forces are balanced the total work performed in the virtual displacement is zero.

The lever is in equilibrium, reasoned Bernoulli, because the forces on it are so connected that if a displacement were to happen, while one of them would perform positive work pulling the body in its direction the other one would be obliged to perform an equal amount of negative work opposing the displacement. The same must be true for any number of forces.

This is the general condition of equilibrium perceived by Bernoulli: any body is in equilibrium if the sum of the work done by all the forces acting on the body is zero for any virtual displacement that may be given to the body. The principle of Bernoulli is now called the *principle of virtual work.*

As an example of the advantages of this principle take the strange looking balance in Fig. 6.18. The arms of the balance are free to move up and down. Intuitively we know that the loads at the ends of the arms must be equal for equilibrium and this is easily confirmed by experiment. Intuitively we also feel that if one of the loads is pushed closer to the support the state of equilibrium will be disturbed. What is the magnitude **Q** of the load required for equilibrium if it is applied at point N rather than at the end of the arm?

Let us give the device a virtual displacement by pushing the right arm

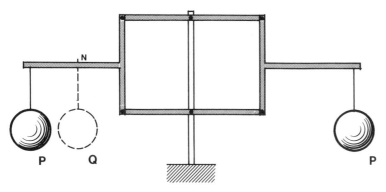

Figure 6.18   Intuitively we know that the loads at the ends of this balance must be equal for equilibrium.

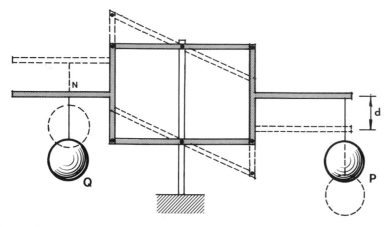

Figure 6.19   Paradoxically, the balance remains in equilibrium even when the loads are at unequal distances from the fulcrum!

down a distance d (Fig. 6.19). The left arm moves up the same distance; both arms remain horizontal. The principle of virtual work states that the total work performed by the forces in a displacement is zero. Therefore

$$P \times d - Q \times d = 0$$

and

$$Q = P$$

Paradoxically, it appears that the load **Q** which is much closer to the fulcrum will maintain the balance in equilibrium if it has the same magnitude as the load **P** acting at the far end of the opposite arm. Furthermore, the respective distances of the loads from the support do not appear in the equation at all.

We must conclude that the forces will balance wherever we may choose to place them on the arms of the balance as long as their magnitudes are equal.

An experiment would confirm this unexpected conclusion. The principle of work supplies the explanation. Note that in any displacement the arms of the balance move the same distance in opposite direction and remain horizontal. Either force performs the same amount of work regardless of its position on the arm of the balance. For one force to do positive work in going down the other force would have to do an equal amount of negative work in going up. The forces are in equilibrium.

We have not discovered anything by means of the principle of virtual work that we may be unable to elucidate in another way, by the principle of moments. It is just that for this particular problem the principle of work provides an answer more quickly and more directly. A specially useful feature of this approach is that we need to consider only the active forces and not the support reactions, which do no work in the displacement.

There are other advantages. To understand our balance we do not even need to know the mechanism that makes it work: it could be enclosed in a box with only the two arms sticking out as in Fig. 6.20—all we need to note is that the two arms move through equal distances and remain parallel when displaced. The requirements of equilibrium can be deduced from that observation alone.

Finally, we should realize that the principle is already familiar in a different form, given to it by Torricelli. Wherever the loads may be on the balance (Fig. 6.21) their common center of gravity G (right in the middle between them when they are equal) stays in the same place as we move the arms up or down. The balance is in equilibrium because the composite body sitting

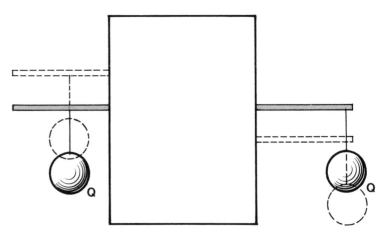

Figure 6.20 The principle of work enables us to analyze the balance even when the mechanism is hidden.

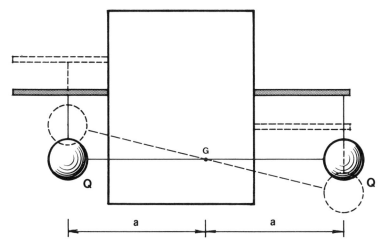

Figure 6.21   The principle of Torricelli does the same and turns out to be just another form of the principle of work.

on it cannot fall lower in a displacement although its separate components (the two equal loads) can move relative to each other.

Torricelli's approach is, in essence, little different from the principle of virtual work. The center of gravity is the point at which the weight force of a body is applied. If the body is supported so that its center of gravity does not move down when a virtual displacement is imparted to the body, the work of the total weight force performed in the displacement is zero. But that precisely is the condition of equilibrium based on virtual work.

STABILITY OF EQUILIBRIUM

The concept of a virtual displacement requires a subtle refinement if the principle based on such displacements is to be of general value. In some cases, if the body is displaced the whole situation may change. For example, the relation between the load **W** and the pull **F** on the inclined plane (Fig. 6.22) is the same for any displacement if the plane is straight. But the slope of the plane may vary from point to point. The force **F** that is required to keep the body in equilibrium at the point where the straight and curved plane have the same slope will be either too large or too small if the body is displaced up or down the curved plane.

We are in a dilemma: a displacement is needed to test the equilibrium of the body yet if one is made the picture is distorted. René Descartes was the first to realize this problem and to suggest an answer. We should not consider a displacement, said Descartes, but only what happens at the instant when the body *begins* to displace. Bernoulli, who had at his disposal more

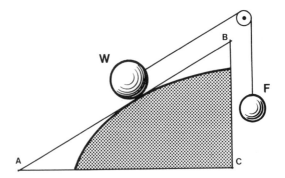

Figure 6.22   The body on the uneven plane: to displace or not to displace?

abstract mathematics, required the virtual displacement to be *infinitely* small. This beginning of motion, the infinitely small displacement, is now accurately defined in modern mathematics.

In a real displacement the work of the forces acting on a body in equilibrium may not be zero. The forces may perform either a positive or a negative amount of work. This gives us additional information about the state of equilibrium of a body.

Suppose we have two identical rods, one supported at the bottom on a

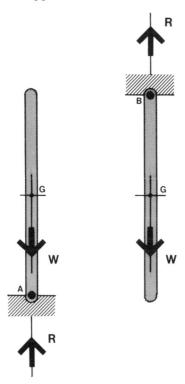

Figure 6.23   Two rods in equilibrium but supported in very different ways.

hinge at point *A* and the other hanging from a hinge at point *B* (Fig. 6.23). There are two forces acting on the body: the weight force **W** and the support reaction **R** in the hinge. If we give either rod an infinitely small displacement—in other words, if we consider what happens when the rod just begins to move—we find that the weight force is displaced in a horizontal direction while the reaction is not displaced at all. No work is done; the body is in equilibrium.

In a real displacement, however, different things happen to the two rods. Let us look first at the rod hanging from point *B* (Fig. 6.24). When the rod is rotated the center of gravity *G* where the weight force acts moves on a curve ending up higher than its original position. Such displacement means negative work and the force opposes it. Therefore when the disturbing influence that caused the displacement is removed, the rod will return to its original position. The body is in *stable equilibrium.*

The rod in Fig. 6.25 behaves differently. In an infinitely small displacement the forces perform no work. They are equal and opposite and therefore in equilibrium. But in a real displacement when the rod is rotated through an angle, the weight force moves again on a curve but this time the vertical displacement of the center of gravity is in the direction of the force. The force performs positive work and (unlike human workers) it will continue to perform all the positive work it can possibly do. In other words, the slightest

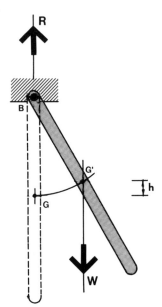

Figure 6.24   Stable equilibrium: after displacement the body returns to its original position.

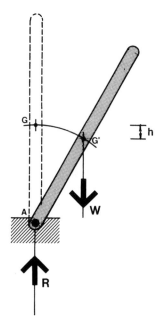

Figure 6.25 Unstable equilibrium: the body falls down when displaced from the position of equilibrium.

disturbance will make the rod fall down. The equilibrium of the rod is *unstable*.

We obtain the same answer if we consider the moments of the forces about the point of support when the body is displaced. In Fig. 6.24 the moment of the weight force tends to rotate the rod back to its original position while in Fig. 6.25 the moment of the same force would cause rotation that would increase the displacement and stop only when the rod is flat on the ground.

Finally, let us support the rod by putting a hinge right at the center of gravity (Fig. 6.26). Now the total work performed by the forces is zero whether we give it an infinitely small or a large rotation since the point at which they both act remains where it was. Whatever the angle through which we turn the body it will remain in the new position, neither trying to return where it was nor continuing to move. From the point of view of moments, both forces pass through the center of gravity and have no moment about it. The body is in *neutral equilibrium*.

Turn the rod until it is horizontal and hang equal loads from its ends. The body—which we can now call a lever—is still in neutral equilibrium. Most simple machines are, by design, in neutral equilibrium and this is why the principle of work (as stated, for example, in Stevin's common rule) can

Example   151

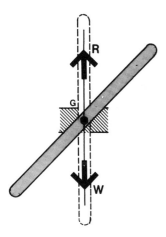

Figure 6.26 Neutral equilibrium: the body remains displaced.

be applied to them in real displacements. The total work performed by the forces on the machine in neutral equilibrium is zero in any displacement, small or large.

EXAMPLE

The penny-farthing bicycle of the 1870s (Fig. 6.27) worked on the principle of the wheel-and-axle machine in reverse: the effort was applied to the small-radius pedal lever to turn the large front wheel. The advantage was that a small displacement of the pedal produced a large displacement of the bicycle and therefore speed. As this bicycle developed, the front wheel tended to get bigger and bigger for more speed. In one design, the wheel was so big that the rider had to sit inside the wheel to pedal.

Instead of increasing the size of the wheel, modern bicycles (Fig. 6.28) reduce the size of the "axle" while still keeping the radius of the pedal lever a comfortable size. The pedals are attached to a pair of gear wheels called sprockets. A cluster of such sprockets is attached to the rear wheel. In the top gear, the chain connects the larger of the two front sprockets with the smallest of the rear sprockets. Suppose the front sprocket has three times as many gear teeth as the rear one; then for each full revolution of the front sprocket the rear wheel is forced to make three.

The principle of work tells us that the force applied to the pedal must come out at the other end of the machine as a push on the ground reduced in proportion to the increased distance of travel. Because of this considerable reduction it is difficult to start up in top gear and almost impossible to

Figure 6.27 The penny-farthing bicycle. The front wheel had to be large to achieve speed.

Figure 6.28 Modern bicycles have the means to change the ratio of force to distance traversed as required.

Example   153

climb a hill. The 10-speed bicycle therefore provides 10 different gear ratios to suit every occasion. For steep hills we use the lowest gear in which the chain is on the smaller of the two front sprockets and on the largest one of the rear cluster.

CHAPTER SEVEN

Summary and Problems

Or statics in a nutshell

THE RIGID BODY

*The lever in Fig. 7.1 is similar to a sketch in one of the notebooks of Leonardo da Vinci. A load of 10 newtons hangs on the right arm of the lever while an unknown load **P** is attached to the left arm by means of a string that passes over a pulley under the fulcrum of the lever. How much load do we need to keep the lever balanced? Since the load is directly under the fulcrum, can it at all balance the other load?*

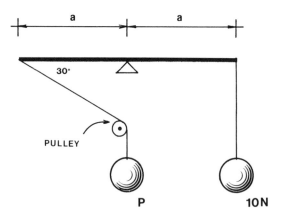

Figure 7.1 *Problem 1:* How much load **P** do we need to keep the lever balanced? Can it be done if the load hangs right under the fulcrum?

The lever is a rigid body in contact with other bodies—the loads and the support. The first step is to draw a free-body diagram showing the body we are investigating and the forces acting on it (Fig. 7.2). The lever is first isolated from all other bodies. This makes it a free body. Next, the bodies removed in the process of isolation are replaced by the forces they exert on the free body.

The load at the end of the right arm of the lever is a force of 10 newtons pulling down. The body is also in contact with the fulcrum; there must be a force at that point too. We do not know its direction and magnitude. Still, we have to show the force. The direction of the arrow is purely a guess, as indicated by the squiggle on the shaft of the arrow. Its magnitude is given as R. In the process of isolating the free body we detached the string at the left end of the lever. That string exerts a pull on the body in the direction of the string, its magnitude equal to P. The weight of the lever itself is neglected in this problem. The information giving the location of the forces (the lengths of the lever arms, the angle of the string) completes the free-body diagram.

The free-body diagram settles part of the question right away. Can force **P** balance the load even though the pulley and the load **P** itself are directly under the fulcrum? The location of the pulley is obviously irrelevant: it is not even part of the free-body diagram. All that counts is the pull of the string at the point where it is attached to the body. The pulley is just a red herring.

The free body is in equilibrium, therefore the forces acting on it satisfy the conditions of equilibrium. We have here a special case: the lever is a three-force body. The three forces have a zero resultant moment, therefore they must meet at a point (Fig. 7.3). Two of the forces (the two loads) have known directions and their lines of action meet at point J. The third force, the unknown one acting at the fulcrum, must also pass through that point and this fixes its direction. The three forces have a zero resultant force and

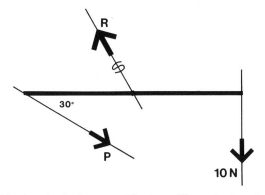

Figure 7.2   The free-body diagram of the lever. The pulley was just a red herring.

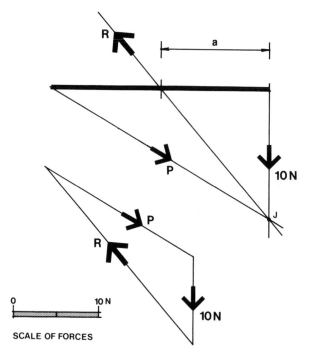

Figure 7.3   The solution, method 1.

therefore form a vector triangle in which they follow each other tip-to-tail, starting with the force of 10 newtons drawn to a convenient scale. The magnitudes of the unknown forces can be scaled off the diagram. Thus we find not only the required force in the string but also the force in the fulcrum.

A more general approach ignores the special circumstance that the lever is a three-force body. Let us again draw the free-body diagram of the lever but this time we resolve all the forces acting on the body into their vertical and horizontal components. The force on the right arm has only a vertical component. The force at the fulcrum is unknown and we therefore show two components, both of them of unknown magnitude. If the force is purely horizontal or purely vertical one of the components will turn out to have zero magnitude. The direction of the arrows is a guess—it will be corrected in the solution if the guess proves wrong. Finally, the components of the pull of the string are shown at the left end of the lever (Fig. 7.4).

Once again we apply the conditions of equilibrium. The sum of the moments about any point must be zero. Take moments about the fulcrum. Three components have no moment about that point because their lines of action pass through the fulcrum so their moment arms are zero. The remaining two forces (the load of 10 newtons and the vertical component of

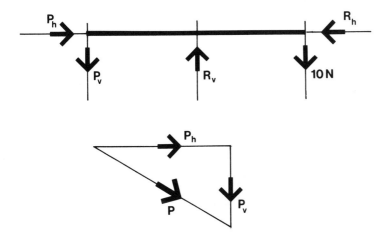

Figure 7.4   The solution, method 2. The horizontal component of the reaction is moved sideways on its line of action for clarity. Its effect is not changed by this.

the string force, P_v) have equal moment arms and therefore must be of equal magnitude (P_v = 10 newtons). Once we know the vertical component of the force in the string we can reconstruct the force itself and its horizontal component from the vector sum

$$\mathbf{P_v} + \mathbf{P_h} = \mathbf{P}$$

seen in Fig. 7.4. We find that P_h = 17 newtons and the pull in the string P = 20 newtons.

The force **R** in the fulcrum can now be found from the remaining condition of equilibrium which states that the resultant force on the body must be zero. If the resultant force is zero, then its components must also be zero. There are two vertical forces of 10 newtons each pulling down at the ends of the lever; for equilibrium, the reaction at the fulcrum must have a vertical component of 20 newtons pushing up (R_v = 20 newtons). There is a horizontal force acting on the body at its left end; this must be balanced by an equal and opposite horizontal reaction at the support (R_h = 17 newtons). Thus we obtain the components of the support force at the fulcrum and from those we can easily find the direction and the magnitude of the force **R** itself.

THE COMPOSITE BODY

The Dutch canal bridge in Fig. 7.5 is designed so that it can be lifted by one person pulling on the rope hanging from the overhanging beam of the bridge.

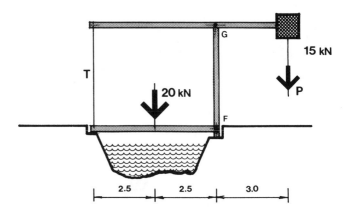

Figure 7.5 *Problem 2:* The Dutch canal bridge. What is the pull **P** required to lift the bridge?

*The weight of the deck of the bridge is 20 kilonewtons and we can assume that it acts in the middle of the deck. The weight of the other members of the structure and friction will be neglected. A counterweight of 15 kilonewtons is attached to the overhang. What is the pull in the rope **P** required just to begin lifting the bridge? What is the tension in the rope **T**?*

The bridge consists of several connected bodies exerting forces on one another. First we draw a free-body diagram of the entire structure. This lets us concentrate on the essential features of the problem, leaving out internal details (Fig. 7.6). What happens inside the free body, between the component bodies, makes no difference to the equilibrium of the body we have isolated since all internal forces appear in pairs. For example, the rope **T** pulls up on the deck of the bridge and it pulls down on the top beam—the net effect on the bridge as a whole is zero. The free-body diagram shows only external forces: the known pull of gravity on the deck and on the counterweight and the unknown pull **P** on the rope.

When the bridge is just raised from its horizontal position it is supported entirely on the pivot at point *F*. The free body is balanced about that point and is in no way (except shape) different from a lever *AB* with a fulcrum at *N*, shown below. We are looking for an external force, the pull in the rope **P**. The condition of rotational equilibrium requires that the sum of all moments about any point be zero. Taking moments about point *F* (or N)

$$-20 \times 2.5 + (15 + P) \times 3 = 0$$
$$P = 1.67 \text{ kilonewtons}$$

This answers the first question.

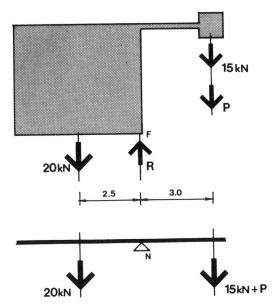

Figure 7.6   The free-body diagram of the entire body.

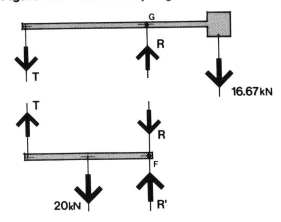

Figure 7.7   Only when the body is cut up into components can we find the internal forces between those components.

The second question requires more work. The force in the rope **T** cannot be detected in the free-body diagram of the entire structure. It is an internal force. We have to isolate one of these two component bodies of the structure on which it acts and then we shall be able to treat the pull of the rope as an external force. Shown in Fig. 7.7 are the free-body diagrams of the deck and of the top beam. Take first the deck. We do not even have to write down the conditions of equilibrium to find the rope force **T**. It is obvious that, since its moment arm to the fulcrum is twice the arm of the load, the pull

$$T = 10 \text{ kilonewtons}$$

We can check this answer by considering the equilibrium of the top beam of the bridge, isolated as a free body. Taking moments about the fulcrum of this lever we have

$$-10 \times 5 + 15 \times 3 + 1.67 \times 3 = 0$$

The body is in equilibrium, as it should be.

The free-body diagram of the top beam enables us to find the support reaction **R** at point G. This follows from the condition that the resultant force on the body must be zero, or, since all the forces on the body are vertical, the sum of vertical forces must be zero:

$$-10 - 15 - 1.67 + R = 0$$

where forces pulling vertically down are given a negative sign, as opposed to those pushing up which get a positive sign. The magnitude of the reaction is $R = 26.67$ kilonewtons. This is the force exerted by the post on the beam. The beam reacts with an equal and opposite force on the post, putting the post in compression. In practice, there would be two posts providing support for the top beam and two ropes holding up the deck. This means that the forces just calculated would be divided in half.

Many practical considerations, very important in actually building the bridge, are completely beyond the scope of our mathematical analysis. For example, we can see in Fig. 7.8 that the heavy framework of the bridge is raised by means of a capstan—an application of the lever. Given sufficient information, we can calculate the moments the workers exert when pushing on the bars of the capstan. But how can we account for the fact (which did not escape the observant artist who drew the picture) that the workers operating the capstan have to jump over the rope *twice* for every complete turn of the machine?

THE INTERNAL FORCES

SALVIATI: If one wished to break a stick by holding it with one hand at each end and applying his knee at the middle, would the same force be required to break it in the same manner if the knee were applied, not at the middle, but at some point nearer to one end?

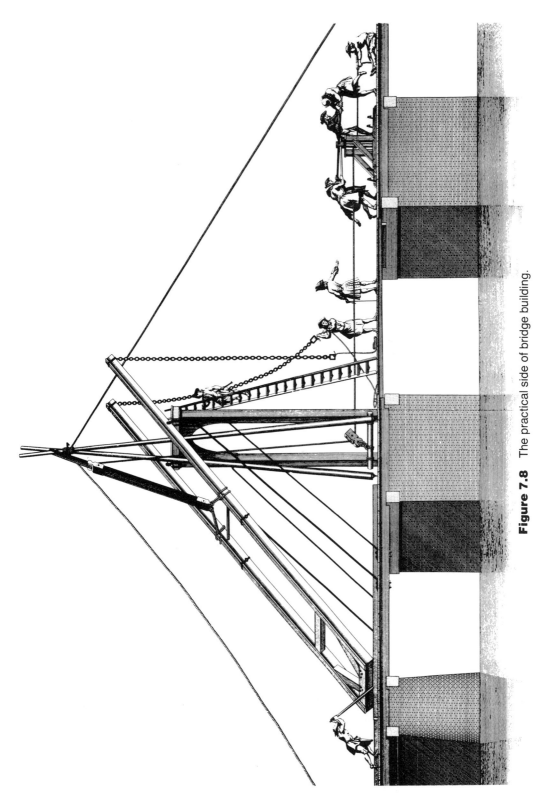

Figure 7.8 The practical side of bridge building.

SAGREDO: At first glance it would appear to be so, because the two lever arms exert, in a certain way, the same moment, seeing that as one grows shorter the other grows correspondingly longer.

This dialogue is taken from Galileo's book TWO NEW SCIENCES. Is Sagredo right?

Let us transform Galileo's dialogue into a structural problem. The stick corresponds to a *beam* on two supports with a load on it (Fig. 7.9). The load takes the place of the push of the knee while the two supports of the beam (opposing the load) have the same function as the pull of the hands at the two ends of the stick. Sagredo believes that the *bending moment* under the load remains the same for any position of the load on the beam: if the load approaches support *B*, the lever arm of the *support reaction* at *B* becomes shorter but the arm of the reaction at *A* becomes that much longer, and vice versa.

It takes Galileo two pages of clever reasoning to prove that this is not correct. We can do the same quickly, without really stopping to think. The procedure is always the same.

First, we have to find the external forces acting on the structure. The beam is in equilibrium and the conditions of equilibrium are satisfied. So if we take moments about point *B*

$$P \times L - F \times b = 0$$
$$P = F \times b/L$$

we obtain the reaction at support *A* for any position of the load on the beam. Taking moments about point *A*, we find that the reaction at the other end is

$$Q = F \times a/L$$

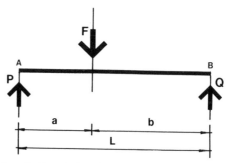

Figure 7.9   *Problem 3:* How to break a stick or avoid breaking a beam.

The reactions change as the load is moved on the beam (as the distances a and b change). When it is exactly in the center of the span, then $a = b$ and the reactions

$$P = Q = F/2$$

The bending moment under the load changes too with the position of the load. It is an internal reaction and, to find out something about it, we have to make an imaginary cut through the beam and isolate a portion of it from the rest. In a way, the beam is treated as a composite body consisting of so many components glued together to reveal the internal forces between them. Since we are looking for the bending moment under the load, let us cut the beam just to the left of the load and make the left-hand portion a free body. In Fig. 7.10 we see the free-body diagram of the isolated portion drawn to a larger scale for clarity. The body is in equilibrium, so there must be a shearing force acting at the section

$$V = P$$

to satisfy the condition that there is no net vertical force acting on the body. The other condition of equilibrium requires that the moment of the couple created by the reaction **P** and the shearing force **V** be resisted by a counter-clockwise couple at the cut—the bending moment. Therefore the bending moment

$$M = P \times a$$

or

$$M = (F \times b/L) \times a = F \times ab/L$$

when we substitute the expression $(F \times b/L)$ for the reaction P. We now have a formula that gives us the bending moment under the load wherever it may be placed on the beam. The couple consists of a compressive force at the top of the beam cross section and a tension on the bottom.

The formula settles the argument between Sagredo and Salviati. True, as we move the load on the beam the moment arm of one reaction becomes shorter in proportion as the other one becomes longer. This does not mean that the bending moments under the load remain the same. They change from a minimum when $a = 0$ or $b = 0$ (that is, when the load is directly on top of a support) to a maximum when $a = b = L/2$. At that point, when the

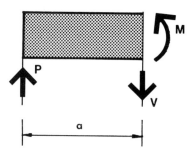

Figure 7.10 The free-body diagram of a portion of the beam, showing internal forces.

load is at mid-span, the bending moment has its maximum value

$$M = F \times L/4$$

Salviati, Galileo's mouthpiece in the dialogue, arrives at the same conclusion but he uses geometry rather than algebra in his discussion. Fully convinced by Salviati's geometrical demonstrations, Sagredo exclaims: "Must we not confess that geometry is the most powerful of all instruments for sharpening the wit and training the mind to think correctly? Was not Plato perfectly right when he wished that his pupils should be first of all well grounded in mathematics?"

HYDROSTATICS

Let us borrow from Stevin a balance with a container partly filled with water taking the place of the regular pan on one side (Fig. 7.11). The weight of the water is found to be 10 newtons while the weight of the container itself is negligible. Next, a lump of lead hanging from a thread is lowered into the water until it rests on the bottom of the container. Suppose the balance now requires a load of 43 newtons in the other pan for equilibrium. What will be the reading if we lift the lump of lead a little (by pulling on the thread) so that it is still completely submerged in water but no longer touches the bottom? The density of lead is 11 times that of water.

Isolate the lump of lead as a free body (Fig. 7-12). The balance reading tells us that the lump of lead weighs 33 newtons. This force is opposed by the pull of the thread and the *force of buoyancy*. The force of buoyancy is equal to the weight of the water displaced by the submerged body and, since lead is 11 times denser than water, we have

$$B = 33/11 = 3 \text{ newtons}$$

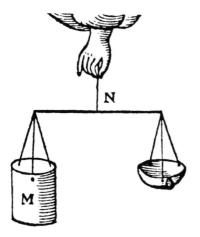

Figure 7.11 *Problem 4:* experiments with a water container on a balance.

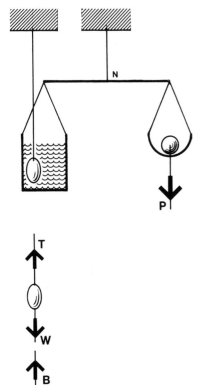

Figure 7.12 A lump of lead immersed in water and supported on a string.

The pull of the thread must supply the remaining 30 newtons required for equilibrium.

Buoyancy is the force exerted by the water on the submerged body. The body exerts an equal and opposite force on the water. Thus an extra 3 newtons pushes down on the container and the balance will now require 13 newtons of counterweight in the pan on the other side. The total force required to hold up the lump of lead and the container with water is still equal to 43 newtons: the string carries 30 and the balance 13 newtons.

This simple problem makes it easier to accept an experiment suggested by Stevin to illustrate the *hydrostatic paradox* (Fig. 7.13). The apparent paradox is in the circumstance that the pressure of a fluid depends only on the depth of the fluid and not at all on the amount of it in the container.

The container on the balance is filled to the top with water and it is found that its weight is 100 newtons. In other words, a counterweight of 100 newtons must be placed in the pan on the opposite side of the balance to preserve equilibrium. The solid cylindrical block, held up by the rigid support as shown in the diagram, is now pushed into the container so that most of the water escapes with only a small amount remaining on the bottom of the container

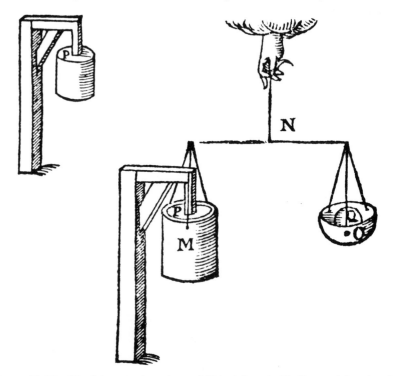

Figure 7.13 Stevin's experiment: a solid block inserted in the container leaving only a small amount of water in it.

and around the cylinder. The remaining water weighs, let us say, only 10 newtons. What counterweight is required now for equilibrium?

The depth of the water in the container has not changed, says Stevin, therefore the pressure on the bottom of the container (which depends only on the depth and not on the amount of the fluid) also remains unchanged. It follows logically that we still require 100 newtons on the other side of the balance to keep the much reduced weight of water in equilibrium. This is good logic but may be a bit difficult to accept intuitively.

Another look at the problem with the lump of lead should help overcome this difficulty. The balance is no longer supporting just a mass of water. The cylinder is buoyed *up* with a force equal to the weight of the water displaced by it, which is 90 newtons; the same force pushes *down* on the container in addition to the weight of the water.

Why did we require more counterweight in the first example with the lump of lead hanging in the water than without it? Remember that in that case we did not let any water escape from the container. The water level in it rose after the lead was lowered inside. In fact, the water level rose by exactly the volume of the immersed body; the additional pressure on the bottom of the container was the same as if a corresponding quantity of water had been added to that already inside.

In the second case, the container was filled to the top and all the excess water was allowed to pour out (as in the lift lock discussed in Chapter 5). The pressure on the bottom did not change. Finally, note that if the water were to freeze, buoyancy would disappear and 10 newtons would suffice to hold up the solidified water.

THE METHOD OF WORK

*The device in Fig. 7.14 is similar to the common scissor-type car jack and, in various forms, is part of many machines. The two bars FD and FE at the top are hinged to the bars DB and EA of length 2L. These bars are connected in the middle, at point C, by a hinge. When our jack is squeezed at the bottom by a pair equal and opposite forces **Q** it assumes the shape shown by the dotted lines and lifts the load **P** at the top. What is the ratio of effort **Q** and load **P**?*

Let us first see how we would solve this problem by the methods reviewed so far. Figure 7.14 can serve as a free-body diagram of the device. We conclude from it that the vertical reactions at points *A* and *B* must each be equal to *P*/2. This diagram tells us nothing about the equal and opposite forces **Q:**

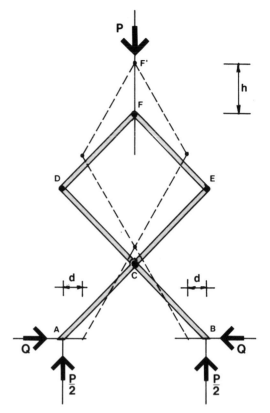

Figure 7.14 *Problem 5:* The scissors-type jack. What are the forces required to lift a given load **P**?

they could have any magnitude without affecting the equilibrium of the free body in any way.

We proceed by breaking up this composite body into components. We start at the top and isolate joint F as a free body (Fig. 7.15). The accompanying vector triangle shows the forces acting on the joint. Identical forces must be exerted at points D and E on the bars supporting the joint. We resolve them into their vertical and horizontal components.

The forces acting on the bar AE (also isolated as a free body) at joint E are equal and opposite to those that our bar exerts on the bar above. Note that we cannot proceed by the *method of joints* because the bars forming the cross press against each other at point C. Thus they have a third force acting on them, unlike the bars FD and FE which either pull or push at their end joints, acting either in tension or compression. The third force means that our bar AE bends and does not exert forces along its length.

Bar AE is in equilibrium. Taking moments about point C we have

$$3 \times P/2 \times a - Q \times a = 0$$

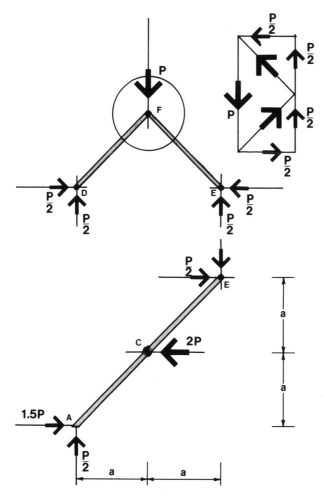

Figure 7.15   Free-body diagrams of the component bodies.

and

$$Q = 1.5\,P$$

The force exerted on the bar at joint C must be purely horizontal or the sum of vertical forces would not be zero. To satisfy the condition of horizontal equilibrium, it must push to the left with a magnitude equal to $2P$.

In this problem the *method of virtual work* will quickly show its advantages. The *work* of a force is defined as the product of the magnitude of the force and the displacement in the direction of the force. In principle, the total work of the forces is zero in a displacement. Therefore, if we push the two forces **Q** forward a distance d and the load rises through a distance h, we have

$$2 \times (Q \times d) - P \times h = 0$$

and

$$Q = P/2 \times h/d.$$

All we have to do is find the ratio h/d. Then the force **Q** required to lift the load can be found directly.

The work of the load is negative, since the displacement h is against the direction of the force. The vertical reactions in the supports A and B are not considered—there is no displacement in the vertical direction so they perform no work. All internal hinge forces that we laboriously discovered through separate free body diagrams are left out—they come in equal and opposite pairs and always the positive work of one of the pair is canceled by the negative work of its partner. If the machine had many more bars, the economy of this method would be even more impressive.

There is just one hitch. For this method we have to displace points A and B but the moment we do that the shape of the jack changes and the forces we find are no longer those that act in the original configuration. Just look at joint F: the angle between the bars changed when the load was raised through a distance h and the vector triangle would now give us a different value for the bar forces.

The way out of this dilemma is, of course, a *virtual displacement*, an imaginary one that is also infinitely small. Of course, we cannot show such a displacement in the diagram. Therefore, the rise h must be calculated. We save time by not having to isolate all those free bodies, but now we have to do some mathematical work with infinitely small displacements. This is done by simple mathematics and is not shown here.

If the displacement d of points A and B was infinitely small, we would find that joint C goes up by the same amount d. Vertical displacements accumulate as we go up: joints D and E are raised by $2d$ while the top moves up by $h = 3d$. Hence

$$2Q \times d = P \times 3d$$

and

$$Q = 1.5\,P$$

as before. The method of virtual work does not produce different results, only new insights. We do not have to go through all the calculations to gain a qualitative understanding of the jack. A glance at the deformed shape of the device when the squeeze is applied (dashed line in Fig. 7.14) tells us immediately that here force is not multiplied: the load is displaced more than the effort. This jack raises the load fast, but at the cost of greater effort.

STABILITY

The center of gravity of a chain suspended between two supports at the same level is at a point equally distant from both supports because of symmetry. We cannot guess where exactly this point is located on the axis of symmetry but it is not on the chain itself. Suppose now one of the supports is lowered so that no axis of symmetry exists (Fig. 7.16). Is the center of gravity C now closer to the support A or B?

Consider the chain as a free body. Each support exerts a pull on the chain, represented by the forces **P** and **Q.** These must be in line with the chain at

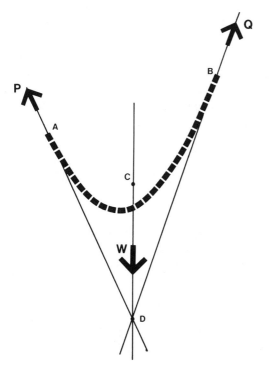

Figure 7.16 *Problem 6:* A chain hanging from two supports is a three-force body. The three forces must intersect at a point.

points *A* and *B* (they are tangents to the curve of the chain at those points, to be precise). There is another force acting on the body: the pull of gravity, a vertical force **W** passing through point *C*. The chain is therefore a three-force body and the three forces must meet at a point. The lines of action of the forces **P** and **Q** are given and they meet at point *D*. The weight force must pass through the same point. We still do not know the vertical location of the center of gravity, but we can see that it is closer to the support *B*.

We know the directions of the two reactions because we know the shape of the chain—but how do we know what shape the chain will take? *Ernst Mach* in his book *The Science of Mechanics* relates the following story: "John and James Bernoulli, on the occasion of a conversation on mathematical topics during a walk in Basel, lighted on the question of what form a chain would take that was freely suspended and fastened at both ends. They soon and easily agreed in the view that the chain would assume that form of equilibrium at which its center of gravity lay in the lowest possible position."

The chain is in *stable* equilibrium. The test of stability is to see what happens when the equilibrium of a body is disturbed in some way. If the body returns to its original position after the disturbing influence is removed, the equilibrium is stable. If it remains in the new position, the equilibrium is *neutral*. If the body continues to displace under the action of the forces acting on it without any other external influence, it is in *unstable* equilibrium.

We can test the stability of our chain very simply by giving it a pull—a virtual displacement (Fig. 7.17). The shape of the chain will change but it will return to its original shape as soon as the disturbing pull is removed and this tells us that its equilibrium is stable. Even though we pulled *down* on

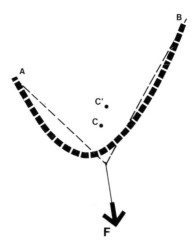

Figure 7.17 Pulling down actually raises the chain. The chain is in stable equilibrium.

the chain, the center of gravity of the distorted shape was actually *raised* by the pull. The change of shape imposed on the chain makes the weight force acting on the body perform negative work in opposing the displacement of its center of gravity.

Mach goes on to say: "As a matter of fact we really do perceive that equilibrium subsists when all the links of the chain have sunk as low as possible, when none can sink lower without raising an equivalent mass in consequence of the connections equally high or higher. When the center of gravity has sunk as low as it possibly can sink, when all has happened that can happen, stable equilibrium exists. The *physical* part of the problem is disposed of by this consideration. The determination of the curve that has the lowest center of gravity for a given length between the two points A and B is simply a *mathematical* problem."

The curve assumed by a chain hanging under its own weight is called a *catenary* and is, indeed, fully defined in mathematical textbooks.

Finally, Mach makes a comment that can be applied to everything that was discussed in this book: "Let it be remarked in conclusion that the principle of virtual work, like every general principle, brings with it, by the insight which it furnishes, disillusionment as well as elucidation. It brings with it disillusionment to the extent that we recognize in it facts which were long before known and even instinctively perceived, although our present recognition is more distinct and more definite; and elucidation, in that it enables us to see everywhere throughout the most complicated relations, the same simple facts."

Bibliography

Ernst Mach, *The Science of Mechanics*, Open Court Publishing Co., LaSalle, 1960.

J. L. Meriam, *Statics*, John Wiley & Sons, New York, 1966.

D'Arcy Thompson, *On Growth and Form*, Cambridge University Press, Cambridge, 1971. (abridged edition)

Blaise Pascal, *The Equilibrium of Liquids and the Weight of the Mass of the Air*, Columbia University Press, New York, 1937.

Simon Stevin, *Principal Works*, Vol. 1: *Mechanics*, Edited by E. J. Dijksterhuis, C. V. Swets & Zeitlinger, Amsterdam, 1955.

Pierre Duhem, *Les Origines de la Statique*, A. Hermann, Paris, 1905.

Galileo Galilei, *Dialogues concerning Two New Sciences*, Dover Publications, New York, 1954. New translation by Stillman Drake, University of Wisconsin Press, Madison, 1974.

Friedrich Klemm, *A History of Western Technology*, The MIT Press, Cambridge, 1964.

Martin Jensen, *Engineering around 1700*, Danish Technical Press, 1969.

Derrick Beckett, *Bridges*, Paul Hamlyn, London, 1969.

M. Salvadori & R. Heller, *Structure in Architecture*, Prentice-Hall, Englewood Cliffs, N.J., 1963

E. J. Dijksterhuis, *Simon Stevin: Science in the Netherlands around 1600*, Martinus Nijhoff, The Hague, 1970.

Index